WAU ECOLOGY INSTITUTE HANDBOOK No. 11

MEDICINAL PLANTS OF PAPUA NEW GUINEA

Part 1: Morobe Province

Edited by Ellen Woodley

The publication of this book was kindly supported by **Bread for the World,** Stuttgart. Their valuable contribution is acknowledged with thanks.

CIP-Titelaufnahme der Deutschen Bibliothek

Medicinal plants of Papua New Guinea / ed. by Ellen Woodley. – Weikersheim : Margraf.
NE: Woodley, Ellen [Hrsg.]

Pt. 1. Morobe province / [contributing authors: F. Goeltenboth ...]. – 1991
(Wau Ecology Institute handbook ; No. 11)
ISBN 3-8236-1185-2 (Margraf)
ISBN 9980-73-002-1 (Wau Ecology Inst.)
NE: Goeltenboth, Friedhelm; Wau Ecology Institute: Wau Ecology Institute ...

Contributing authors:
F. Goeltenboth, D. K. Holdworth, H. Sakulas, H. Thredgold, E. Woodley

Illustrations:
Shaw, Powell, Nolau, Pearce, Fourner, Volau, Suraga, Edwards

Printing and binding:
F. & T. Müllerbader
Germany

Mühlstraße 9, P. O. Box 105
D-6992 Weikersheim, Germany

ISBN 3-8236-1185-2
ISSN 0935-9109

Copublished by:
Wau Ecology Institute, 1991
P.O. Box 77
Wau, Papua New Guinea

National Library Papua New Guinea
ISBN 9980-73-002-1

ACKNOWLEDGEMENTS

There are many who have contributed to this book in one way or another and while we would like to mention each person individually, this is neither practical or possible. First and foremost, we wish to thank all of the village people encountered over the years who patiently and willingly shared this important knowledge of theirs. We gratefully acknowledge the financial assistance from the Morobe Provincial Government, the Papua New Guinea Biological Foundation and the Office of Environment and Conservation. A lot of time and effort was given by the staff at the National Herbarium for the identification of numerous plant specimens and for the use of some of their botanical illustrations. Thanks also go to the staff and students of Dregerhafen High School who contributed to a survey with David Holdsworth in their home villages along with their elderly relatives. Also to Mr. Pano Manaha and Mr. B. Boze for work done in the Huon Peninsula supported by the P.N.G. Biological Foundation. We would also like to thank Mr. Artis Venus and Mr. Aubeta Kairo at the P.N.G. Forestry College in Bulolo for their time spent in the identification of plant specimens. Thanks are also extended to Mr. Jim Croft for his time and support given at the early stages of this book and Dr. Klaus Dobat, Botanical Garden University of Tuebingen, for his support. Finally, without the hard work and sensitivity of Mr. Lawong Balun, who collected together with Ellen Woodley for the Wau Ecology Institute, the *Medicinal Plants of Morobe Province* would not have gone this far.

D.K.H.
H.S.
E.W.

THE WAU ECOLOGY INSTITUTE

P. O. BOX 77

WAU

PAPUA NEW GUINEA

The Wau Ecology Institute is located in the mid-mountain area at Wau (Morobe) in Papua New Guinea.

The Institute was established in 1972 as a non-profit, non-governmental corporation dedicated to the study and education of ecology and conservation, rural development and environment protection in Papua New Guinea.

Located at about 1000 meters altitude on the lower slopes of Mount Kaindi, the Institute has access to a remarkable range of environments at various altitudes.

The summit of Mt. Kaindi (2350 m) is accessible by road, and the extensive pristine forests provide abundant wilderness tracts for survey and study of the biota.

The Institute has facilities for short-term and long-term visitors. Short-term guests may stay in the hostel or in rental bungalows. Long-term researchers and visitors may lease Institute housing and laboratory space. In addition, the Institute supports a regional research library, arboretum, herbarium, small zoo, medicinal plants demonstration garden and preserved collections of vertebrates and invertebrates. Research facilities include dehumified laboratory space, microscopes, preserving equipment and assorted research supplies.

Visitors, researchers, naturalists and bird-watchers are welcome to visit and stay at the Institute. Because of limited space, all are adviced to make bookings in advance, and to make full inquiries about the availability of specific services, supplies or transportation:

The Director
Wau Ecology Institute
P. O. Box 77
Wau, Papua New Guinea
Phone 44-63 41

AUTHORS AND THEIR CONTRIBUTIONS

Present address

Dr. Friedhelm Goeltenboth
Universitas Kristen Satya Wacana
Il. Diponegoro 52 – 60
Salatiga 50 711
Indonesia
(Distribution and comparison of collected plants in Morobe Province)

Dr. D. K. Holdworth
Chemistry Department
Universiti Burnei
Burnei 3186
(Preface; Medicinal plants of Morobe Province)

Harry Sakulas MSc.
Wau Ecology Institute
P. O. Box 77
Wau
Papua New Guinea
(Preface; Medicinal plants of Morobe Province)

Dr. Harald Thredgold
University of Technology
Private Mail Box
Lae
Papua New Guinea
(Medicinal plants of Morobe Province)

Ellen Woodley	125 Ferguson St. Guelp Ontario NIE2Y9 Canada (Introduction; Mcdicinal plants of Morobe Province; Appendix; References; Index)
Andrew Yauieb	Ambassador of Papua New Guinea Gotenstraße 163 5300 Bonn Germany (Foreword)

FOREWORD

In Papua New Guinea our way of life is closely integrated with the plant and animal life of the country. Our predecessors depended upon the richness of the forests, oceans and rivers for their daily food, clothing, building materials, tools, weapons and medicines. Over the years from the days of discovery through colonial rule to present day, new life styles were introduced and adopted by our people. However, in the village-life we were never far removed from the natural resources which continued to supply many of our needs.

Over the years the economic value of our natural resources has been recognized. Forests resources for sawn timber, plywood, wood chips and the wealth of our grasslands for grazing by animals are the most apparent avenues of development. To effectively develop and utilize these natural resources for economic, social or medical benefits, demands that we have detailed knowledge of the kind of plants be they trees, herbs, creepers, etc, growing in our country.

This handbook describes the traditional medicinal uses of over one hundred species of plants currently used in Morobe Province.

Due to many constraints, it is not possible to cover everything in the Province and it is unlikely that this will even be possible. However, a cross section of geographical and cultural areas has been covered.

It is indeed pleasing that D. K. Holdsworth, E. Woodley, H. Thredgold, F. Goeltenboth, and H. Sakulas have been able to complete the organization of the text and illustration for the Wau Ecology Institute Handbook No. 11 of Traditional Medicinal Plants of Morobe Province, Papua New Guinea. We hope that the guidelines and models established by the authors will provide the basis for publication of similar handbooks for different Provinces. This is essential for the development of new medicine from Papua New Guinea's traditional medicine as much as there has been from European, Asia, North America and African plants.

I would like to compliment the authors and all those who have assisted in preparing this volume. Also appreciation is extended to E. Woodley who has completed and edited the Handbook.

I extend my gratitude to all concerned in the initiation of this project that will immensely benefit Papua New Guinea.

Andrew M. D. Yauieb, CMG

The Ambassador of Papua New Guinea
to the Federal Republic of Germany

PREFACE

This handbook describes the traditional medicinal uses of over one hundred species of plants currently used in Morobe Province.

Due to many constraints, it is not possible to cover every village in the province and it is unlikely that this will ever be possible. However, a cross section of geographical and cultural areas were covered during the period from 1978 to 1985, notably:

1. The Finschhafen coast
2. Mountains of the eastern Huon Peninsula
3. The Snake River Valley, near Buang and Mt. Shungol
4. Aseki, in the area behind the Etuti dividing range
5. The Upper Watut Valley behind Bulolo
6. Buso, on the Morobe coast
7. Wau area

All of the plants collected were identified at the National Herbarium, Office of Forests in Lae and the scientific names are used throughout the book. Local names are also provided and pertain to the villages where the collection was made. Where possible, a careful search of the literature was made to find uses of the plant in other areas of Papua New Guinea and other countries. At the University of Papua New Guinea, references were checked directly from data bases in Australia and the United States. In many cases, plants which grow in different parts of the world are found to be used independently for similar purposes.

The book is intended to be of assistance to health workers and administrators, responsible for extending the coverage of health services. They can develop a fruitful working relationship with practitioners of traditional medicine who live and work with and who enjoy the confidence of the majority of rural people.

It is envisaged that more areas in the province will be surveyed in the near future and that this book will be enlarged and revised every few years. Similar books for other provinces may be possible if there is sufficient interest and financial assistance.

A more comprehensive book of the medicinal plants is being written. It is based on the book *Traditional Medicinal Plants of Papua New Guinea* by D. K. Holdsworth, published by the South Pacific Commission in 1977 and long since out of print. It will contain information on over three times as many plants as in the first edition. Over four hundred different species representing over one hundred plant families are included.

With the introduction of hospitals and first aid centres, many people do not use traditional medicines as they would have twenty years or more ago. It has been noticed that many children and adolescents who would have formerly been initiated into the customs of their society now do not receive initiation. Often they have to leave their village to go to school or to join relatives who have immigrated to towns. As a consequence, the traditional uses of plants are not often passed on to the younger generation. As older people die, much traditional wisdom and knowledge is lost.

We have endeavoured to collect information that might otherwise have been lost forever. Foremost, it is essential that the continued use of these plants be encouraged. As well, it is possible that useful new medicines may be developed from Papua New Guinean traditional medicine, as there have been from European, Asian, American and African plants.

David K. Holdsworth
University of Papua New Guinea
Port Moresby

Harry Sakulas
Wau Ecology Institute
Wau, Morobe Province

CONTENTS

LIST OF FIGURES

Introduction

This multi-authored handbook represents the result of an integrated effort to document and describe plant species that are used in traditional medicine in villages of Morobe Province, Papua New Guinea. The final compilation is 127 species of plants from 20 villages.

What this book does not cover is an account of all of the medicinal plant species in use throughout the province. Funding is limited, as is the time and human resources required to undertake a project of that scope. The terrain is rugged and access to the more remote villages is possible either by small aircraft or many days journey by foot. It is also necessary to spend time with people in the village, as it is very important and often guarded information that is being conveyed to an outsider. It would be insensitive to hastily put pen to paper before regarding the feelings of the traditional healer or person who posesses the knowledge. All of these pose considerable constraints and leave the option of investigating only a sample of the villages within the province. In spite of the limitations, a great deal of information has been compiled for Morobe Province and it is hoped that this handbook will be the stepping stone for revised editions of even greater scope.

It is believed by all of those involved in the handbook that this accumulation of knowledge that has been handed down through the generations is a rich heritage and must continue to be preserved through succeeding generations. The continued use of medicinal plants must be encouraged. Attempts are being made at the Wau Ecology Institute to structure a program whereby the people in villages and health care workers recognize the value of the indigenous medical system and do not wholly convert to and rely upon an imported, modern medical system. Where appropriate, the traditional system should be appreciated both medically and socially and integrated into a structure where the traditional and modern systems work side by side. A medicinal plant garden has been initiated at the Institute to cultivate as many medicinal plants as possible with the aim to demonstrate their value and to provide the incentive for villages to propagate their own most useful medicinal plants. Another aspect of the ongoing work at WEI is the beginning of a chemical testing laboratory where tests for alkaloids, flavanoids, saponins, steroids, tannins and triterpenoids may be conducted.

The handbook presents a list of the plant species along with their botanical description, their medicinal use and their chemical properties. The text is quite technical and the specific information about each plant species may be of interest to those who chose to attempt an identification of the plant and learn more of the plant in terms of its structure and origin. This information is a starting point – a reference.

The body of the main text of the handbook gives detailed information on individual medicinal plants and is divided into sections.

Each species is listed by scientific name under its family name, both of which are in alphabetic order. On the same line is the collection number for the specimen if it is filed at the Wau Ecology Institute Herbarium. Voucher specimens are held at the National Herbarium and duplicates are at WEI Herbarium. Under the species name is the Greek or Latin derivation of that name, which is included for interest.

The first section for each species describes the plant itself. The first paragraph within this section for each species is a **botanical description** of that species, which is of general interest and may be of use if one wishes to identify a plant with those in the handbook. In this regard, this description compliments the illustrations.

The second paragraph is a brief account of the **ecology** of the species; its habitat, conditions in which it grows and the kind of environment it may be found in.

The third paragraph is the **distribution** of the species, by country or region. The origin of the species is included if this is known. The next section is entitled **"Medicinal use"** and is what the book is all about. Within this section:

The first paragraph includes the use(s) of that species in Morobe Province only. This information is from the collections of David Holdsworth, Lawong Balun and Ellen Woodley.

The second paragraph describes the use(s) of that species in the rest of Papua New Guinea. This information is almost entirely from the work of Holdsworth, unless referenced otherwise. In this paragraph, reference is frequently made to the name of a province where a particular species is used. This in no way implies its use throughout the province. Instead, it is to group similar uses together in a province and to avoid the repeated use of village names in a text that is intended only to summarize. For a complete list of village names and local plant names for those species found outside Morobe Province, reference must be made to the book *Traditional Medicinal Plants of Papua New Guinea* by D.K. Holdsworth, 1977.

The last paragraph includes the use(s) of the species in other countries in the world.

The references to a plant's uses in Morobe Province alone would not paint a complete picture. The inclusion of other areas in Papua New Guinea and the world gives further insight into the extent the plant is used and thereby lends credence to the medicinal value of the plant. It is interesting to read that one species may be used extensively within Morobe Province, but another thing to learn that it is used all over PNG and in a number of other countries as well; such is the case with *Alstonia scholaris* or *Wedelia biflora*, for example. Conversely, it is equally noteworthy when a species is of use in one particular village and then recorded nowhere else, as in the case of *Epipremnum pinnatum* or *Tecomanthe dendrophila*.

The following section, entitled **"Chemistry"** briefly outlines, where known, the important chemical constituents of that species. This was done by means of a literature search, by Harold Thredgold at the University of Technology in Lae. More insight into how a medicinal plant is effective physiologically or psycologically may be gained from this section. It is this kind of information that preceeds further exploration by pharmaceutical companies seeking drug-yielding plants.

Lastly, the **local name(s)** of the plant are listed below the scientific name, along with the **name of the village** in Morobe Province, where it is used.

The illustrations in the handbook are from a variety of sources. Twenty were drawn by Sally Shaw, during her tenure as a U.S. Peace Corps volunteer in 1983 – 85. The others were borrowed from the National Herbarium in Lae.

The work behind the written word involves the integrated effort of many in addition to the direct contributors to the handbook. Travel over rugged terrain was by 4-wheel drive, boat, light aircraft but ultimately ended up on foot. This often required the help of people to guide the way through dense bush to a more remote village, to provide food and a place to sleep. People were always ready and willing to help, always friendly and interested in what was going on and there was rarely a problem in having a knowledgable person convey the cherished wisdom of traditional medicine. It is these people that must be thanked and left with the hope that the knowledge of traditional medicine will remain and flourish amidst the flood of external influences that are present in Papua New Guinea today.

Distribution and comparison of collected plants in Morobe Province

The areas and villages where collections were made are shown in Figure 1. The Finschhafen district was the area collected in most intensively. There is a remarkably large number of medicinal plants used in this area today. Figure 2 illustrates the number of different plant species found used in each district of Morobe Province. This reflects the amount of sampling done in each area. More intensive sampling throughout the province is required to determine how much variation in the use of traditional medicine there is from area to area. In the Finschhafen District (F), 83 species were collected, which represents over half of the number collected for the entire province. In Mumeng District (M), 23 collections were made, in Wau District (W), 21 species were recorded, in Menyamya District (MY) there were 35 and in Lae (L) District, only 8 species were collected. No collections have yet been made from the Districts of Kabwum (KA) and Kaiapit (K).

28 of the collected plants were screened for their active ingredients and 18 of these were found to contain alkaloids and 14 of these are considered poisonous. Further investigations are underway, in cooperation with pharmaceutical and chemical institutes in P.N.G. and West Germany, to determine chemical properties of these plants.

The 127 plant species recorded are grouped into 71 plant families. This represents 58 % of the 122 plant families with medicinal properties found for the entire country (Holdsworth, 1985 pers. comm.; Webb, 1960).

The medicinal plants collected were found to be used in 70 different treatments. Table 1 lists the ailments recorded and the number of plant species used in the treatment of the ailment. For example, 39 different species are used to treat sores, bruises and swellings; 26 species are used in the treatment of malaria and fever and another 26 species are used to treat cuts and wounds. Some plants in particular are used to treat a very broad spectrum of diseases, for example *Alstonia scholaris* which is used in 14 different treatments.

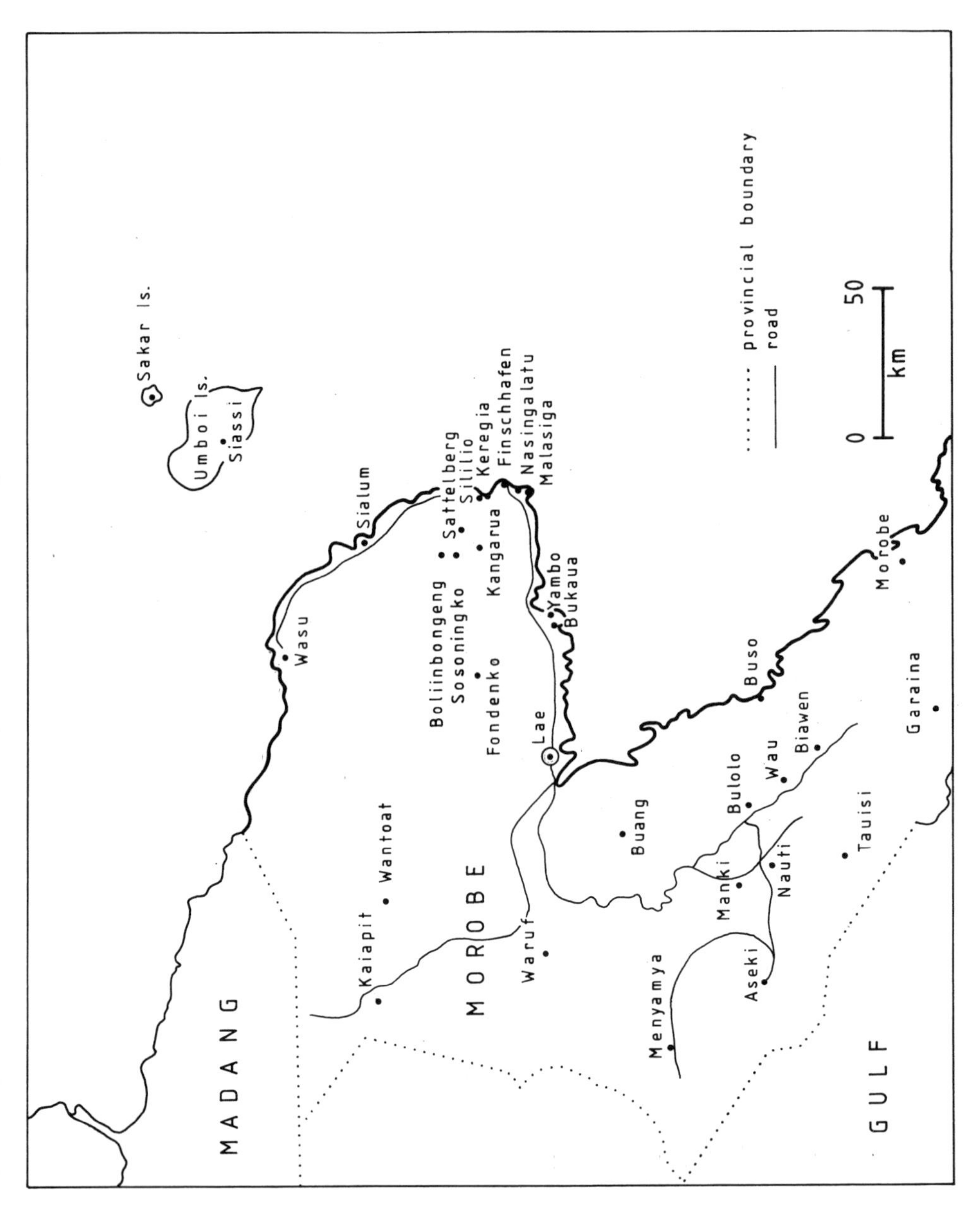

Fig. 1 Map of Morobe Province showing the study areas where plant collections were made (•).

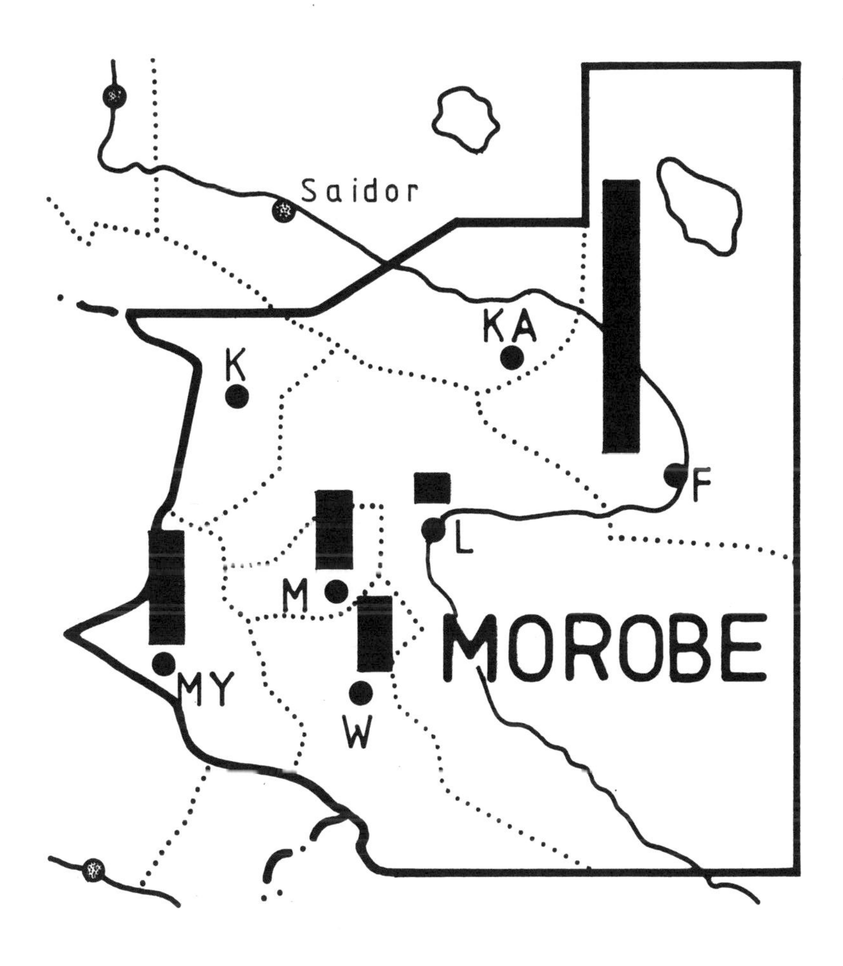

Fig. 2 Distribution of collected medicinal plant species by district in Morobe Province.
F Finschhafen; **K** Kaiapit; **KA** Kabwum; **L** Lae; **M** Mumeng; **MY** Menyamya; **W** Wau.
The bars indicate the absolute number of different plants collected in each district.

Table 1 Numbers (n) of medicinal plant species used in specific treatments.

Treatment	n	Treatment	n
Sores/Bruises/Swellings	39	Induction of Fertility	4
Stomache Ache/Gastritis	30	Snake Bites	4
Skin Rash/Scabies/Pain	29	Internal Sores/Cancer	4
Diarrhoea/Dysentery	26	Hookworm	4
Malaria/Fever	26	Insect Bite/Urchin Sting	3
Sore Throat/Cough	26	Constipation	3
Cuts/Wounds	26	Fatigue	3
Boils/Abcesses/Ulcers	22	Weak Bladder	3
Headaches	16	Induction of Sterility	2
Colds/T.B./Pneumonia	16	Head Lice/Fleas	2
Toothache	10	Menstrual Problems	2
Bronchitis/Asthma	10	Hepatitus/Jaundice	2
Skin diseaes	10	Broken Bones/Dislocation	2
Tonic	8	Baby Ailments	2
Sedative	8	Gout	2
Sore Eyes	6	Sore Breasts	2
Swollen Stomach/Groin	6	Whooping Cough	2
Abortifacients	6	Anaemia	2
Earache	6	Cataracts	2
Poison Antidote/Purgative	6	Rheumatism	2
Pregnancy Pains/Labour	5	Indigestion	2
Sore Mouth/Tongue	5	Heart Condition	1
Rashes/Pimples	5	Convulsions	1
Venereal Diseases	4	Chest Complaints	1
Swollen Spleen	4	Bleeding	1
Vomiting	4	Urinary Problems	1
Contraceptives	4	Senility	1

Medicinal plants of Morobe Province

ACANTHACEAE

<u>*Graptophyllum pictum* (L.) Griff.</u> (Fig. 3)

(Gr. graphe – picture; phyllon – leaf)

Description. Shrub, 2–3 m tall. Leaves opposite, oblong, acuminate; may be green, silver, yellow, red or variegated. Flowers axillary or in terminal racemes. Corolla 2 lipped, dark red. Fruit a cylindrical capsule.

Ecology. Common in primary forest.

Distribution. Pantropical. Widespread at middle to low altitudes in PNG.

Medicinal use. In two villages in the mountains near Finschhafen, the strong stems are used as splints for broken bones and the liquid from the scraped bark is massaged into the skin around the break. In Buang, the sap is used on spear wounds to enable the extraction of the spear head. This shrub is also used in Indonesia (Perry 1980) and Malaysia (Burkill 1966), where the leaves are applied to chronic ulcers and sores, and in the Philippines, where the bruised leaves are applied to abcessed gums (Perry 1980).

Chemistry. Alkaloids are not present (Hartley, 1973; Holdsworth, 1983).

Local names	Village names
Nepec	Bolinbaneng
Nepec	Sosoningko
Kutung	Buang

AMARILLIDACEAE

<u>*Crinum asiaticum* L.</u> (Fig. 4)

Description. Large herb. Flower solitary, showy, pinkish-white on a tall robust peduncle. Leaves linear, basal.

Fig. 3 *Graptophyllum pictum* (L.) Griff.
A Habit diagram, **B** Inflorescence.

Fig. 4 *Crinum asiaticum* L

Ecology. Widespread at low altitudes (in Papua New Guinea), often near the coast, behind beaches, at thc inner margins of mangroves and beside streams.

Distribution. Occurs in Southern Asia, Malesia.

Medicinal use. In a Finschhafen area village, the cut root is cooked in a banana leaf, then cooled and placed on an aching tooth.

In other parts of Papua New Guinea, the plant is used in a variety of ways. In the Trobriands, the stem fibers are used to stop bleeding; on Karkar Island and in Simbu, the latex from the leaves is applied to cuts; in New Ireland, the milky sap from the stem is used for stonefish wounds.

The roots are used in New Caledonia (Rageau 1973), Indonesia (Burkill 1966) and Malaysia in a poultice for wounds, ulcers and swellings. In India (Chopra 1956), the leaves are applied to skin diseases and inflamation. The bulb is used as an emetic in the Philippines and Java.

Chemistry. The active substance is an alkaloid, lycorine (Quisumbing 1951). This alkaloid is common in this family. The biological effects of lycorine are described by Manske (1960). Lycorine has anti-cancer and anti-viral properties (Duke, 1985). Other alkaloids present are crinamine and crinine (Manske, 1960).

Local name	Village name
Faropac	Keregia

ANACARDIACEAE

Mangifera minor Bl. (Fig. 5) WEI-MB-12
Wild Mango

Description. Tree, 18–32 m high; buttressed. Flowers in terminal panicles, yellowish and fragrant; five stamens.

Ecology. Found in lowland primary and secondary forests, sometimes up to 400–750 m, occasionally to 1000–1350 m in lower montane forests. Sometimes cultivated near villages.

Distribution. Occurs in the Solomons and New Britain, New Guinea, Moluccas, Sulawesi, Lesser Sunda Islands, Micronesia.

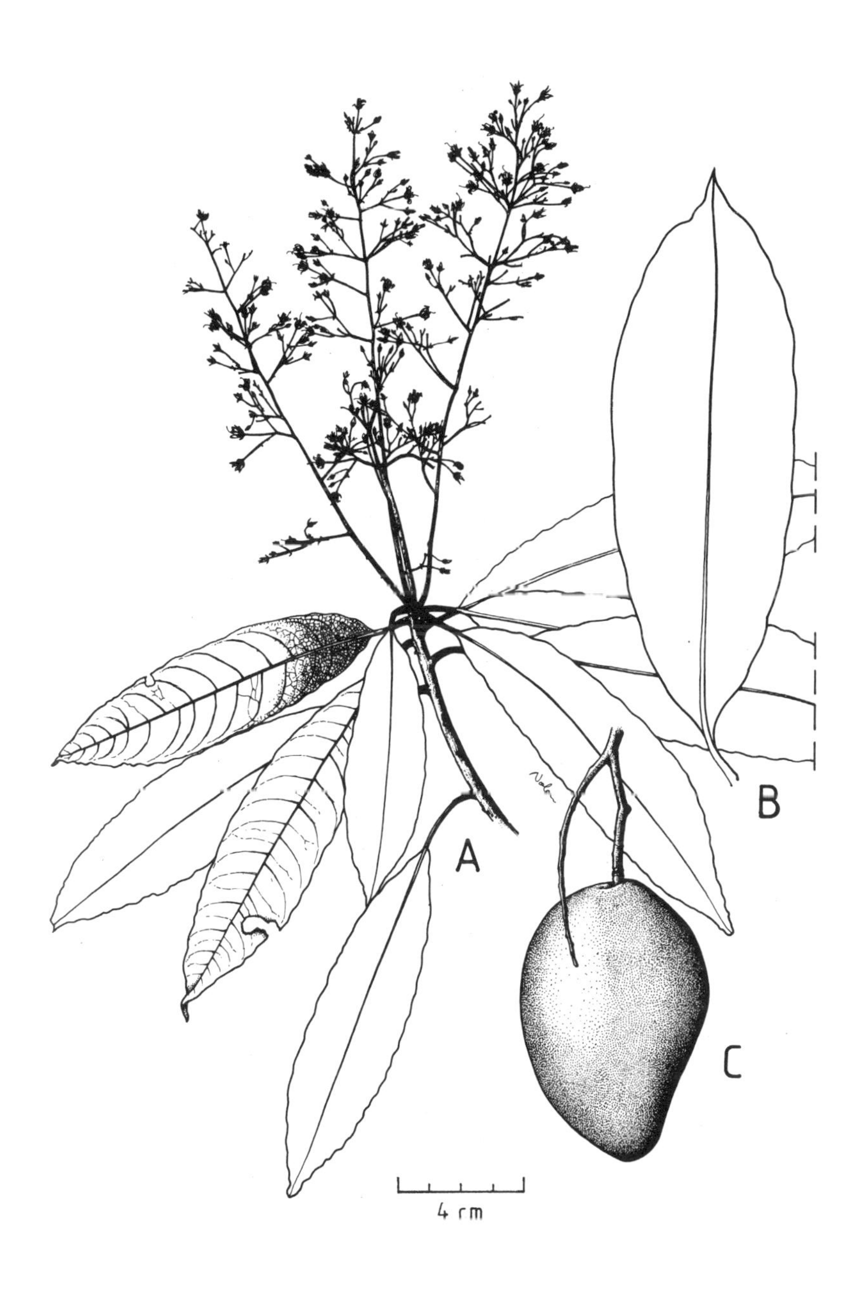

Fig. 5 *Mangifera minor* Bl.
A Habit diagram, **B** Leaf, **C** Fruit.

Medicinal use. In three villages in Morobe Province, the bark is scraped, mixed with water and drunk. The bark may also be chewed to treat snake bite. It is used to treat centipede bite in the Sepik.

Chemistry. Constituents of the cultivated mango (*Mangifera indica*) include a steroid, ambolic acid (Corsano, 1968) and mangiferin, a xanthone (Bhatia, 1967) which is used in pharmaceutical preparations (Chemical Abstracts, 1968).

Local names	Village names
no record	Buso
no record	Mundala
Worang	Keregia

Dracontomelon dao (Blanco) Merr. & Rolfe

Description. Decidious trees up to 43(–55) m tall; buttressed. Leaves petioled, spiral, imparipinnate, with 4–9 pairs of leaflets. Flowers bisexual, in panicles, white or greenish white, five petals, five stamens. Fruit a drupe, brown when ripe.

Ecology. Found only in high rainfall areas, although sometimes in areas with a short dry season. In evergreen to slightly decidious forest on well drained to poorly drained soils, in levee and secondary forest. Common; scattered at low altitude.

Distribution. Occurs in India, Burma, Thailand, Cambodia, S. China, scattered all throughout Malesia and the Solomon Islands.

Medicinal use. A decoction of the leaves is drunk to ease intense side pains. The bark is also of medicinal value (Burkill 1966).

Chemistry. The leaves of the related *Dracontomelon mangiferum* contain an octahydroindoloquinolizine alkaloid (Johns, 1966). Alkaloids are not present in the bark.

Local name	Village name
Loc	Nasingalatu

Rhus taitensis Guill. (Fig. 6)

Description. Tree up to 30 m high. Buttresses sometimes present. Leaves imparipinnate. Flowers in terminal panicles, cream to white, rarely pink. Fruit a drupe; black when ripe.

Fig. 6 *Rhus taitensis* Guill.

Ecology. Found in primary, dryland rainforest, in inundated forest along rivers, sometimes in clearings, secondary forest or savannah. Common in lowland and lower montane zones.

Distribution. Occurs in Polynesia, Micronesia, Solomon Islands, Australia, New Britain and widely distributed in Malaysia.

Medicinal use. To treat severe boils, the bark is scraped and mixed with water and drunk; then the bark residue is placed on the boil and bandaged with a banana leaf. In New Hannover, the scraped bark is mixed with wild ginger and drunk to heal tropical ulcers.

Chemistry. From *Rhus chinensis*, a Chinese herbal medicine effective in the treatment of heart disease and bronchitis is found gallic acid, ethly gallate, a coumarin and several flavones (Gui, Xiao-Ming, 1980). Rhuslactone (a triterpene) has been isolated from *Rhus javonica* (Sung, Chung-Ki, 1980). Flavones and flavanones have been obtained from other species.

Local name	Village name
Qarawec	Keregia

ANNONACEAE

Uvaria rosenbergiana Scheff.

(L. uva – raceme)

Description. Liana, to 30 m high. Leaves alternate, oblong-ovate acuminate, pubescent above. Flowers opposite the leaves in the tree crown, purple-brown. Fruit single, ellipsoid, 12–16 mm long.

Medicinal use. A half a cup of freshly collected stem sap is drunk to relieve fever or malaria. This dose is repeated the next day.

Chemistry. Alkaloids are absent (Holdsworth, 1983). The species *Uvaria ellotiana* contains an indole alkaloid (Achenbach, 1979). Flavanoids and chalcones are present in various species (Lasswell, 1977).

Local name	Village name
Namachaaloc	Zafiruo

APOCYNACEAE

Alstonia scholaris R. Br. (Fig. 7) WEI-S-03, WEI-Q-01
(after the Scottish botanist, C. Alston 1685–1760)

Description. Tree, 30–50 m tall, with white, milky sap. Leaves oblong, arranged in whorls of 4–7, glabrous. Flowers white, scented. Calyx toothed, corolla tube 8 – 10 mm, pubescent outside. Seeds with hairs to 20 mm. The durable wood is used for canoes and carvings.

Ecology. A common lowland species in primary and secondary forest, and in the lower montane rainforest. Also occurs in the monsoon forests and savannah woodland.

Medicinal use. An infusion of the dried bark is used to treat malaria and fever in three Finschhafen area villages. In Buang, the sap is drunk to ease constipation.

Parts of this tree are widely used in Papua New Guinea for malaria. Other uses include: an infusion of the bark for dysentery and diarrhoea on Manus Island and New Ireland and for abdominal pains in the Sepik and Northern Province. The stem sap is mixed with water and drunk to treat cough on Normanby Island and gonorrhoea in Milne Bay. It is used to treat asthma, hypertension, lung cancer and pneumonia at a village hospital on Manus Island.

This plant is used extensively in other areas to treat malaria and fever: in Indonesia, Malaysia and the Philippines (Burkill 1966), Irian Jaya (Warburg 1899). It may also be used to combat diarrhoea in Indonesia (Steenis-Kruseman 1953).

Chemistry. There is a strongly positive reaction for alkaloids; the following have been isolated: ditamine, echitamine, echitenine and echitamidine (Manske, 1965), picrinine (Manske, 1970) and picralinal (Manske, 1973).

Local names	Village names
Zopang	Sililio
Zopang	Suquang
Zopang	Fondengko
Katung	Buang

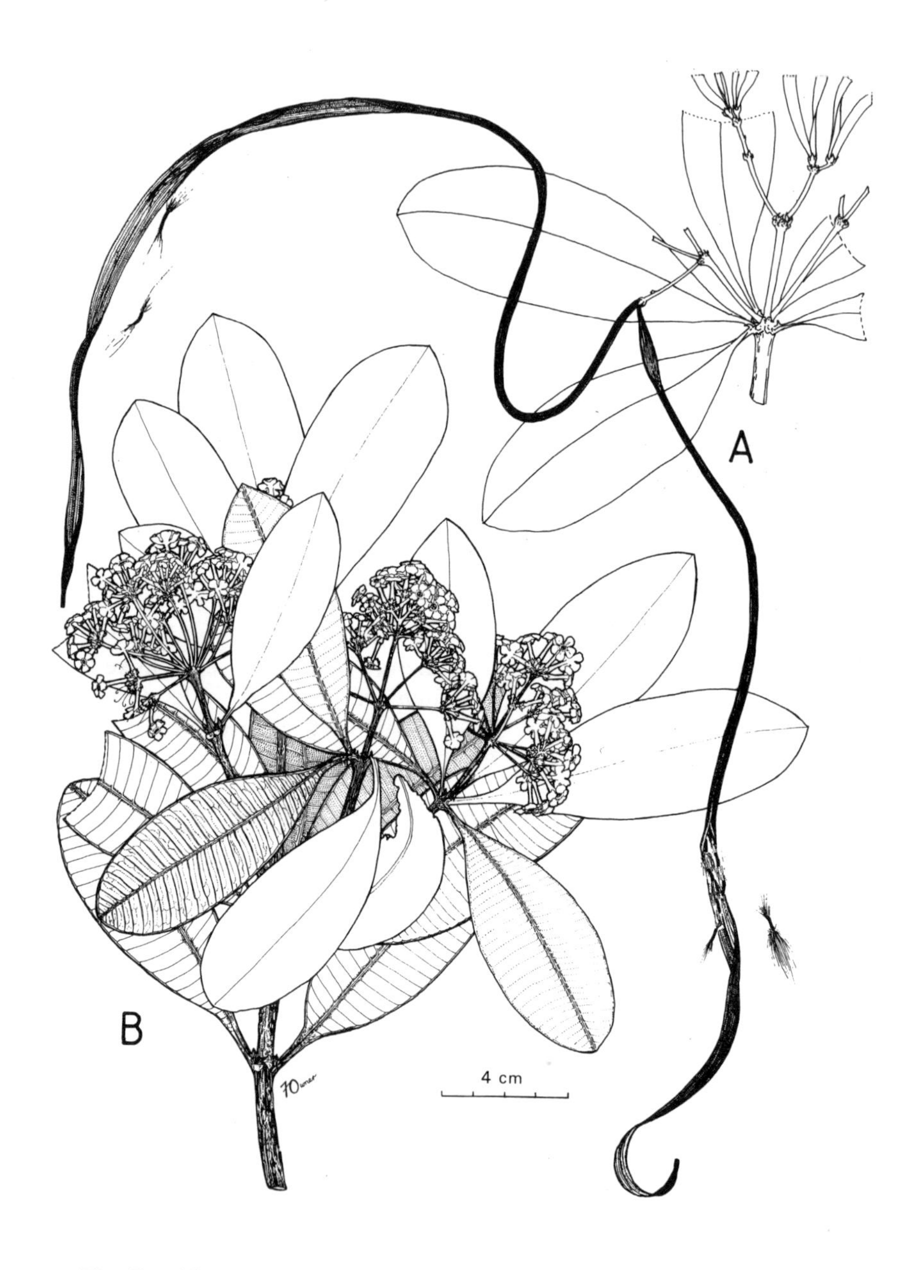

Fig. 7 *Alstonia scholaris* R. Br.
A Seed, **B** Habit diagram

Alstonia spectabilis R. Br. WEI-B-09, WEI-K-02

Description. Tree, 10–20 m tall. Leaves in whorls of 3–4, elliptic to obovate, acuminate. Flowers in many flowered cymes; corolla tube white, 5 mm long. Fruit a follicle, 25–45 cm long.

Ecology. Common.

Medicinal use. In Biawen, the dried bark is crushed and mixed with a small amount of water and drunk to treat malaria, fever and stomach ache. Also, the latex is applied to the hair to kill head lice. Similarly, in Kangarua, the scraped bark is mixed with water and drunk to treat malaria.

In other parts of the country, parts of the tree are mixed with water to treat coughs (Normanby Island and Central Province). In Central Province and Bougainville the bark and stem sap, respectively, are used to treat malaria; in Milne Bay, the sap is applied externally to tropical ulcers; and in Central Province (Brown River) the leaves are used as a poison antidote.

Chemistry. Contains the alkaloids ditamine, echitamine and echitenine. Another alkaloid, alstonamine is also present (Manske, 1965).

Local names	Village names
Silibu	Biawen
Kackacna	Kangarua

Cerbera manghas L. WEI-Bu-116

(after Cerberus, the mythical dog, whose bite was poisonous; manghas, Portugese name for *Mangifera indica*, because of the mango-like fruit)

Description. Tree, 10–20 m tall. All parts of the tree, have a milky-white latex. Leaves alternate, petioled. Flowers have a sweet scent; arranged in terminal cymes, subtended by white bracts. Corolla white; a white star with a red margin in the middle. Fruit a follicle, green, red or purple.

Ecology. Common in primary and secondary forest, on coastlines.

Medicinal use. In the coastal village of Buso, the milky latex from the leaves is mixed with water and rubbed frequently on the skin of children with chicken pox. In New Ireland and Manus Island a related species *C. floribunda* is used as a muscle relaxant, to treat malaria, blackwater fever, hepatitis, gonorrhoea and to assist childbirth.

Chemistry. Alkaloids are not present in the leaves or bark (Hartley, 1973). The seeds are reported to contain cerberin (a heart poison), odollin, sitosterin and myristic acid (Perry, 1980). Cerberic acid and cerberinic acid have also been isolated (Abe, 1977). Cardiac glycosides such as cerberoside, therobioside, deacetyltanghinin and neriifolin are present.

Local name	Village name
Abumung	Buso

Micrechites Miq. sp. WEI-K-04

Description. No information.

Ecology. No information.

Distribution. No information.

Medicinal use. To treat diarrhoea, about a half cup of stem sap is drunk.

Chemistry. Alkaloids are not present in the bark or leaves (Hartley, 1973).

Local name	Village name
Maremu	Kangarua

ARACEAE

Acorus calamus L. WEI-YB-04
(Gr. acorus – a privative; kore – pupil of the eye)

Description. Plant with a thick creeping aromatic rootstock. Stem triangular. Leaves 0.5–1.2 m long. Flowers in a cylindrical spadix, 4–6 cm long, densely covered with flowers. Fruit a berry.

Ecology. A swamp plant, found on the banks of swamps and ponds.

Distribution. Native to Asia; now throughout the world.

Medicinal use. In Yambo, the stem is heated over a fire and the extracted sap is applied to fresh knife or axe wounds.

It is also used widely in other parts of the country: in the Southern Highlands a portion of a dried leaf is eaten with sweet potato to relieve internal sores. In Central Province, the young leaves are crushed and mixed with coconut and heated to make a tonic and a bathing solution. In Enga, the crushed root is rubbed into the hair to kill lice and the leaf is chewed to relieve toothache.

Its importance in other parts of the world is indicated by its reported use in Great Britain as a tonic (Watt et al 1962) and as a tonic in childbirth in Java and Malaysia (Burkill 1966)

In Europe it is used in the manufacture of perfume and liqueurs.

Chemistry. Alkaloids are absent (Hartley, 1973). The essential oil from the root contains asarone, a phenolic ether. Other substances present include calamenol, calamene, calameone, methyleugenol, eugenol and acorin, a bitter component (Duke, 1985). Methyleugenol is a sedative and eugenol is an analgesic and an antiseptic (Duke, 1985).

Local name	**Village name**
Gakoc	Yambo (Bukawa)

Epipremnum pinnatum (L.) Engl.

(Gr. epi – on; premnon – tree trunk)

Description. Climber on trees, 30–50 m high. Leaves acuminate, 60–80 cm long. Flower a cylindrical spadix, 15–20 cm long. Fruit a composite of red berries.

Ecology. Common in rainforest.

Distribution. Occurs in Tanna, New Caledonia, Loyalty Islands, Northeast Australia, Fiji, Tonga, Marshall Islands, Bismarck Archipelago, New Guinea, Moluccas and Malaysia.

Medicinal use. Pieces of the outer stem are mixed with water, then strained and drunk to relieve severe side pains.

Chemistry. No references found. *E. giganteum* is used as one of the components of the dart poison (Perry, 1980).

Local name	Village name
Golong	Keregia

ARALIACEAE

Osmoxylon micranthum Harms. (Fig. 8) WEI-Wau-120

Description. Sparsely branched shrub, to 8 m. Leaves in terminal clusters, petioled. Leaf blade deeply lobed. Inflorescence a terminal compound umbel. Corolla orange or reddish. Fruit an ellipsoid drupe, deep purple or black.

Ecology. Found in primary forest, from the foothills to the montane mossy forest, often in swampy or deeply shaded situations, from 700–2400 m.

Distribution. Occurs only in New Guinea.

Medicinal use. The leaves are chewed with traditional salt to relieve backache and pains in the hip and knee joints. The leaves are also rubbed on centipede bites.

Chemistry. No references found.

Local name	Village name
Diauka	Aseki

Schefflera Forst. sp. WEI-S-09, WEI-F-01

Description. Shrub. Flowers small, in panicles in upper axils. Fruit is subglobose and fleshy.

Ecology. No information.

Distribution. No information.

Medicinal use. In Sosoningko and Fondengko, the stem sap is collected and drunk to treat pneumonia. In Zafiruo, the sap is collected in a piece of bamboo stem and drunk to cure diarrhoea

Fig. 8 *Osmoxylon micranthum* Harms.

and dysentery; in addition, roasted taro (*Colocasia antiquorum*) and sweet potato (kau kau or *Ipomoea batatas*) are recommended to eat with it.

Chemistry. Alkaloids are present in this genus (Holdsworth, 1983). A spermicidal saponin has been isolated from *S. capitata* (Chemical Abstracts, 1978). This acetylenic compound has fungicidal activity (Muir, 1979).

Local name	Village name
Nily	Sosoningko Fondengko Zafiruo

BALSAMINACEAE

Impatiens hawkerii Bull.

Description. Perennial herb, 0.5–1 m tall, stems decumbent to erect, sometimes rooting at the lower nodes, pale to deep green, often tinged with red-purple. Leaves in whorls of 3–7, petioled. Flowers solitary, one to each leaf axil, on upper part of stem, white, pink, lilac, purple, orange, pale red, scarlet, crimson or magenta. Fruit a capsule.

Ecology. Found in moist, shaded or semi-shaded places in montane or sub-montane forests, particularly along stream and river margins, amongst damp rocks or by tracksides. From (200–) 400–3150 m.

Distribution. Occurs in New Guinea and the Bismarck Islands.

Medicinal use. In Nauti, the leaves are rubbed over the abdominal area to relieve pain. In Aseki, the young leaves are chewed with traditional salt to induce labour in pregnant women.

In the Eastern Highlands, the whole plant is cooked and eaten by children with stomach ache and the sap from the fruit and leaves is rubbed onto the legs of children who are slow in learning to walk. Also, some of the Kukukuku people use the leaves to rub the stomach of pregnant women to relieve labour pains (Blackwood, 1935).

Related species of *Impatiens* are used in other parts of the world to treat labour difficulties (China) (Perry 1980). It is an abortifacient in East Africa (Watt et al 1962).

Chemistry. Alkaloids are absent in the leaves and the bark (Hartley, 1973).

Local names	Village names
no record	Nauti
Imda	Aseki

BARRINGTONIACEAE

Barringtonia asiatica L. WEI-YB-01

Description. Tree, 10–20 m tall. Leaves oblong-obovate, entire, sessile, leathery. Flowers large and showy, in racemes, 5–20 flowered, at the ends of the twigs. Corolla white, stamens 10–13 cm long. Fruit 4 angled, corky, 10–14 cm across.

Distribution. No information.

Ecology. Common on beaches.

Medicinal use. In Yambo, the inner bark is crushed and mixed with water and drunk to ease the aching associated with malaria. The dosage is two cups per day for two days.

In other provinces, the fresh nut is scraped and applied to sores. The dried nut is ground into a powder, mixed with water and drunk to cure coughs, influenza, sore throat, bronchitis, diarrhoea and swollen spleen. It is also used in combination with other plants as a medicine to treat tuberculosis in New Ireland and the Solomon Islands.

The seeds are grated and are effective in stunning fish.

Chemistry. Triterpenoid saponins and gallic acid are present (Webb, 1948). Gallic acid is claimed to have bacteriostatic and anti-tumour properties.

Local name	Village name
Puc	Yambo

BEGONIACEAE

Begonia cf. *tafeansis* Merr. & Perry

Description. Succulent herb.

Ecology. No information.

Distribution. No information.

Medicinal use. In Nauti, the leaves are heated and rubbed into the skin to relieve acute abdominal pains.

Chemistry. No information found.

Local name	Village name
Hamawang	Nauti

BIGNONIACEAE

Tecomanthe dendrophila (Bl.) K. Sch. (Fig. 9)
T. giellerupii Steen.

Description. Tall woody liana, to 20–30 m high. Leaves imparipinnate, with 1–2 pairs of leaflets. Flowers in racemes, arrising from the old wood. Corolla tubular; tube pink-red and lobes cream-yellow. Fruit a beaked cylindric capsule.

Ecology. Found in swampy or dry rain forests, sometimes riverine forests, from sea level up to ca. 1500 m.

Distribution. Occurs in Malesia: Moluccas, New Guinea (throughout) and the Solomon Islands.

Medicinal use. Respiratory problems are treated with a half cupful of newly collected bark sap.

Chemistry. Alkaloids are not present in the leaves or the bark (Hartley, 1973 and Holdsworth, 1983).

Local name	Village name
Hahato, Babaneng	Zafiruo

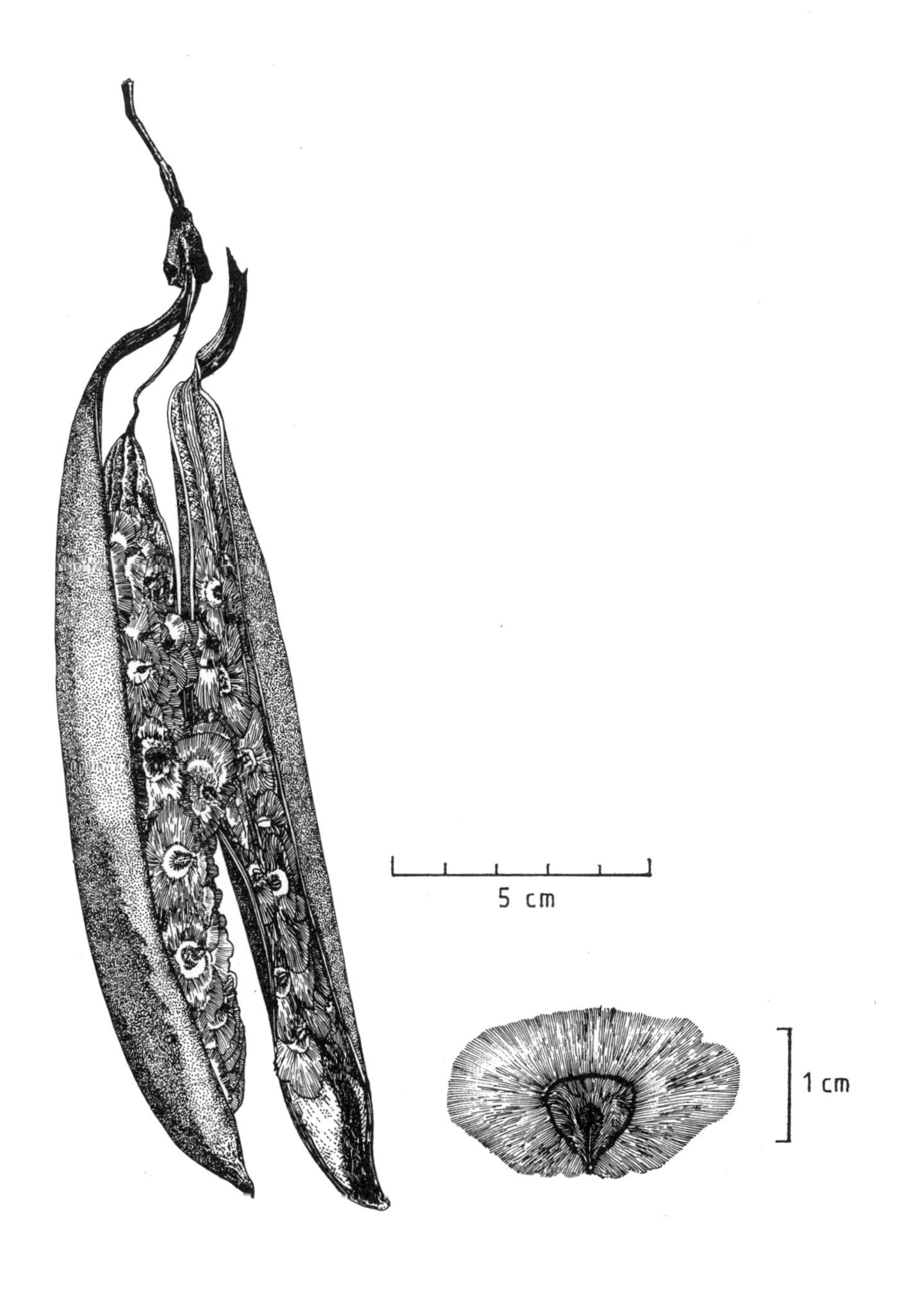

Fig. 9 *Tecomanthe dendrophila* (BL.) K. Sch.

BURSERACEAE

Haplolobus floribundus(K. Sch.) H. J. Lam
(Fig. 10) WEI-Bu-091

Description. Tree, 10–30(–36) m high. Leaves with 1–4 pairs of leaflets. Leaflets oblong. Inflorescence axillary and sometimes terminal. Fruits ovoid.

Ecology. Found in primary forests up to 900 m, where it is relatively common, widely spread and an extremely variable species.

Distribution. Occurs only in New Guinea.

Medicinal use. The young leaf shoots are crushed, then mixed with cold water and drunk to treat stomach ache.

Chemistry. No references found.

Local name	Village name
no record	Buso

CARDIOPTERIDACEAE

Cardiopteris moluccana Bl.

Description. A climbing, twinning herb with white milky latex. Leaves spirally arranged, ovate-cordate. Flowers arranged in panicles.

Ecology. Found in tall rainforest or forest edges, also in secondary vegetation and in native gardens, generally in the lowland, (ascending to 1460 m in New Guinea).

Distribution. Occurs in Malesia: Sulawesi, Philippines, Moluccas, New Guinea and New Britain.

Medicinal use. The stem sap is used to treat asthma and other related respiratory problems.

A decoction of the stem is used against hepatitis in Ternate.

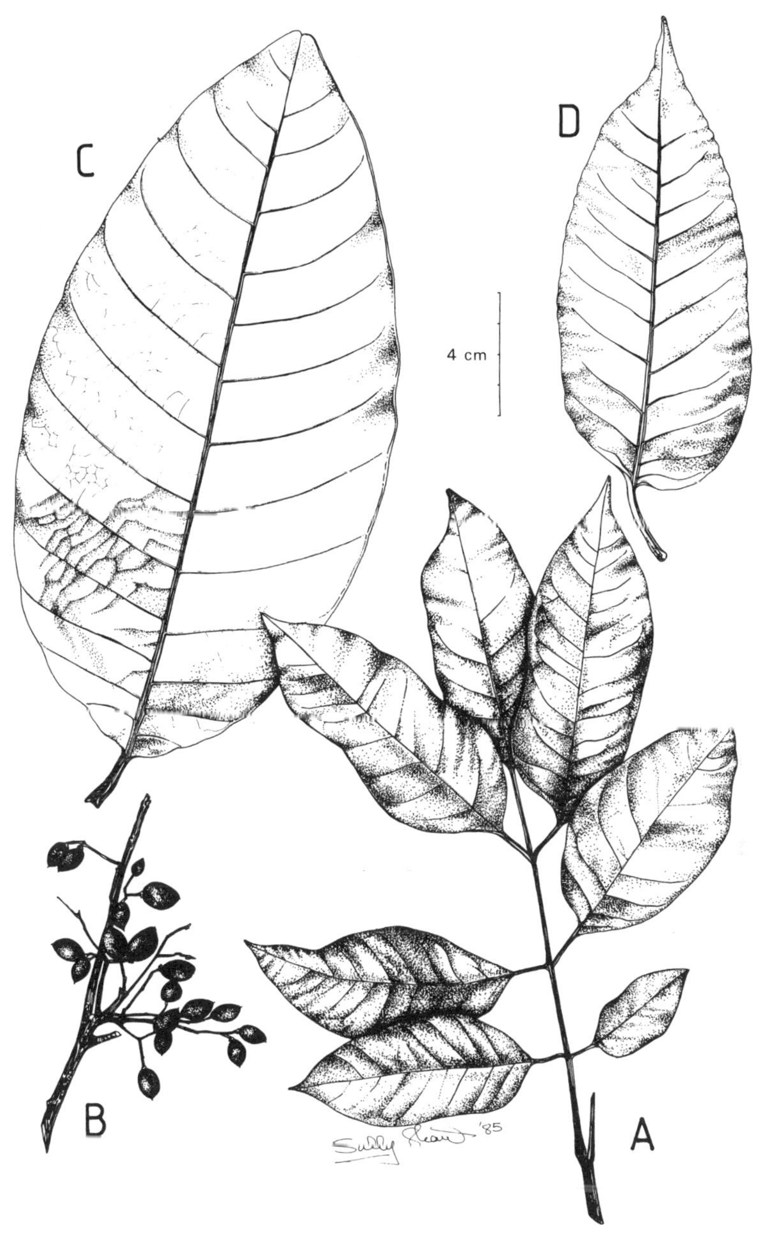

Fig.10 *Haplolobus floribundus* (K. Sch.) H. J. Lam
A Habitat of branch, **B** Inflorescence, **C** Leaf, **D** Leaf.

Chemistry. Alkaloids not present in the leaves or stem (Hartley, 1973; Holdsworth, 1983).

Local name	Village name
Zafengang, Qacac	Sosoningko

CARICACEAE

Carica papaya L.
(Gr. carica-fig)
Papaya, Paw paw

Description. Tree, 4–8 m tall. The leaves are palmately 5–9 lobed and crowded at the stem and branch apicies. Flowers white; the male flowers on axillary panicles and the female flowers solitary or in racemes. Fruit oblong to globose; flesh yellow to pink and edible.

Ecology. A tropical plant, grown in latitudes to 32° N and S. It is killed by frost. It produces good crops near the equator up to 1600 m. Requires full sun, but needs windbreaks. Low temperatures result in fruits of poor flavour.

Distribution. It is never found in the wild, but it is probable that it originated in Southern Mexico and Costa Rica. It may have arisen by hybridization with a closely related species (*C. peltata*). It has now spread to all tropical and subtropical countries.

Medicinal use. In Keregia, the milky juice from the green, unripe fruit is used to treat ringworm. Also, the latex from the leaf stem is blown into the ear to relieve earache.

This tree is widely used in other parts of PNG: the crushed leaves are externally applied to treat headaches, cuts and a swollen groin, the stem latex is used to combat grille (a skin fungus caused by *Tinea imbricata*) and it is ingested as an abortifacient, and is a treatment against diarrhoea.

It is widely used in other parts of the world: it is a vermifuge in Malaysia (Burkill 1966) and the Philippines (Padua et al 1977), it is applied to boils in east Africa (Watt 1962), and used to treat ringworm in the West Indies (Ayensu 1981).

The seeds are used in some countries as a vermifuge, counter irritant and abortifacient.

Chemistry. The leaves contain an alkaloid carpaine (Govindachari, 1965). Carpaine can cause paralysis, numbing of the nerve centres and cardiac depression. The leaves, sap (latex) and the fruit contain the well known proteolytic enzyme papain and chymopapain, both of which have protein digesting and milk clotting properties. Papain has also been used to treat infected wounds.

Local name	Village name
Papae	Keregia

CASUARINACEAE

Casuarina equisitifolia, Forst. (Fig. 11) WEI-BI-121
(Casuarina, from the likeness of the leaves to cassowary feathers)

Description. Tree, with thin branches with scale-like leaves. Leaves small, in whorls of 6, adpressed to the twig. The flowers are unisexual; the male flowers in spikes and the female flowers in cones.

Ecology. A light-demanding species with rapid growth, found on sandy coasts and often planted inland. The roots have nodules containing nitrogen-fixing bacteria.

Distribution. Occurs in the Bay of Bengal and throughout Malesia, to the Pacific Islands.

Medicinal use. In the Finschhafen area, the inner bark is scraped and mixed with water to drink as a cure for dysentery and diarrhoea.

In New Ireland and Central Province, a similar treatment is prepared by washing and scraping the roots. In Enga, the inside of the bark is scraped and the juice is given to sedate a mentally disturbed or aggressive person.

In Malaysia, it is also used as a sedative for demented people (Burkill 1966) and in NE Australia (Queensland) the bark is (was) used as an astringent and to combat chronic dysentery and diarrhoea (Bailey 1909).

Chemistry. No references were found.

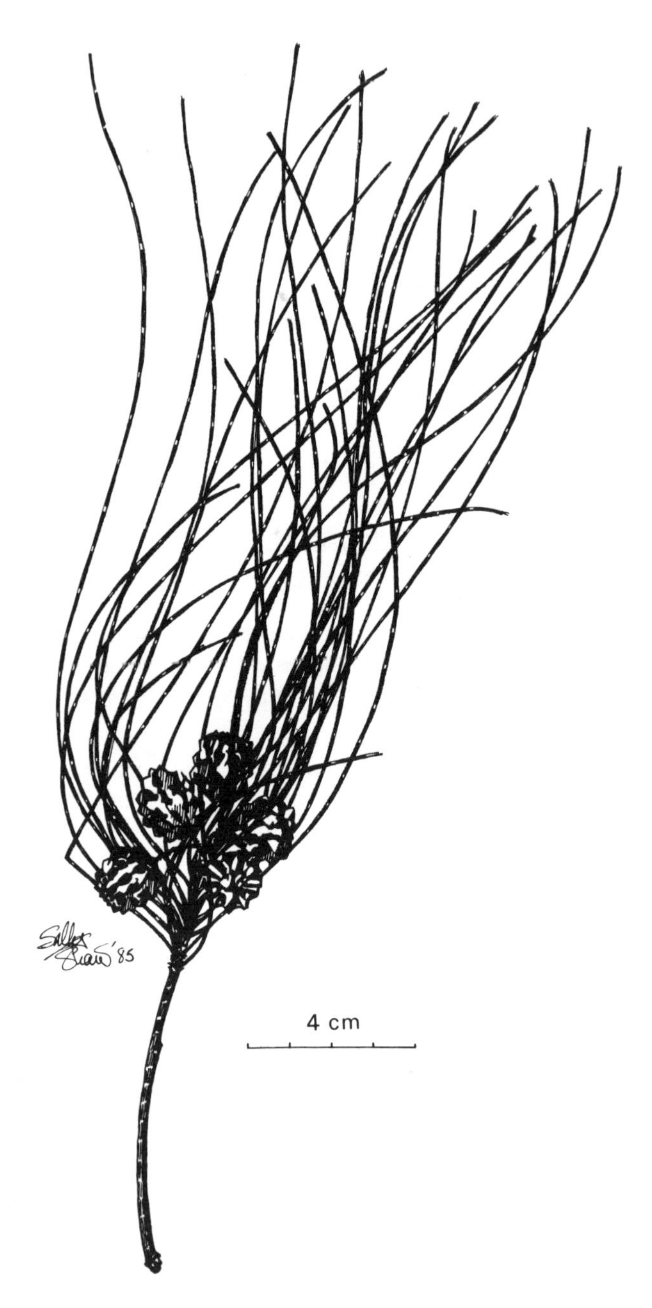

Fig. 11 *Casuarina equisitifolia* Forst.

COMBRETACEAE

Terminalia catappa L. (Fig. 12)
(L. terminus – end; katapa – Malay name)

Description. Tree, 10 – 20 m tall, rarely to 40 m. Leaves obovate, entire, petioled and crowded at twig tips. Flowers in axillary racemes, unisexual, white or cream coloured, small (3–5 mm). Fruit is flattened and usually surrounded by a stiff flange; red when ripe. Leaves turn a brilliant red before falling. The kernal of the fruit is edible; it produces a fatty oil similar to almond oil.

Ecology. A characteristic coastal tree, common on sandy or rocky beaches, usually confined to the tidal zone; also found on river banks.

Distribution. Occurs throughout tropical Asia, northern Australia and Polynesia. It is now widely planted throughout the tropics. In Papuasia, it has been collected from most of the coastal regions.

Medicinal use. In Nasingalatu, the flower is crushed, mixed with water and drunk to induce sterility.

In other parts of the country, uses are varied. On Karkar Island, the sap from the white stem pith is squeezed and drunk to relieve cough and the juice from squeezed leaves is applied to sores. In New Britain, the old yellow leaves are crushed in water and drunk to sooth a sore throat. On Bougainville, the bark is applied to sores and the leaves are heated and placed on pimples (Blackwood 1935).

In other parts of the Pacific, the plant has similar uses: in Somoa, it is used to cure cough and sore throat (Uhe 1964), in Tonga, the juice from pounded bark and leaves is applied to mouth sores (Weiner 1971) and in Irian Jaya, the leaves are applied to wounds and burns (Sterly 1970).

Chemistry. No references for *T. catappa* could be found but several tannins have been isolated from the fruit of *Terminalia chebula* (Haslam, 1967). A triterpene, sericic acid has been isolated from the roots of *T. sericea* (Bombardelli, 1974) and another triterpene, tomentosic acid is a constituent of *T. tomentosa* (Row, 1962; Schneider, 1965).

Local name	Village name
Tali	Nasingalatu

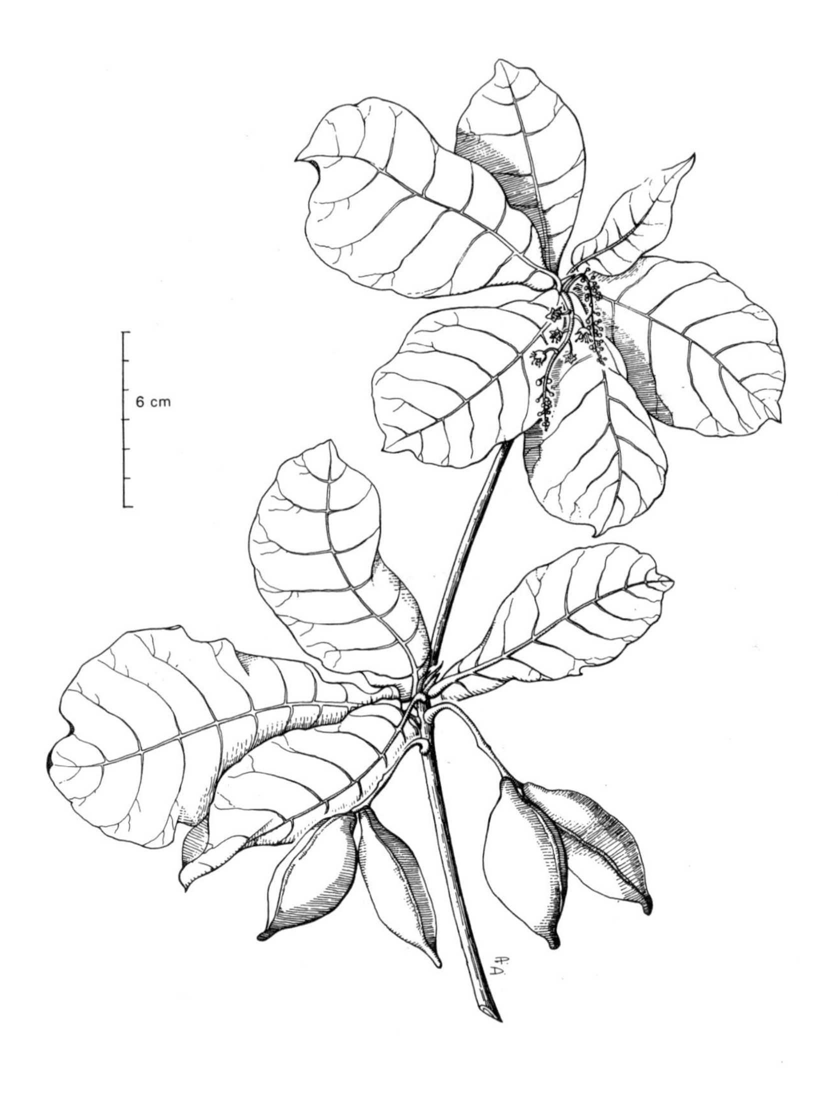

Fig. 12 *Terminalia catappa* L.

COMPOSITAE

Ageratum conyzoides L. (Fig. 13) WEI-Wau-20
(Gr. a geratos – not ageing)

Description. Tall herb, 7–90 cm, with a branched hairy stem and a strong unpleasant smell. Leaves opposite, serrate toothed and petioled. Flowers in terminal corymbs, consisting of 8–15 heads. Corolla pale mauve, lavender, lilac, pale blue. Achenes linear-oblong, black, glabrous.

Ecology. A common weed of roadsides, cultivation, wasteland, plantations and pasture, on floodbanks, in low secondary growth, edges of forest, often very common or abundant. In New Guinea from sea level to 2500 m.

Distribution. Native to tropical America and now pantropical.

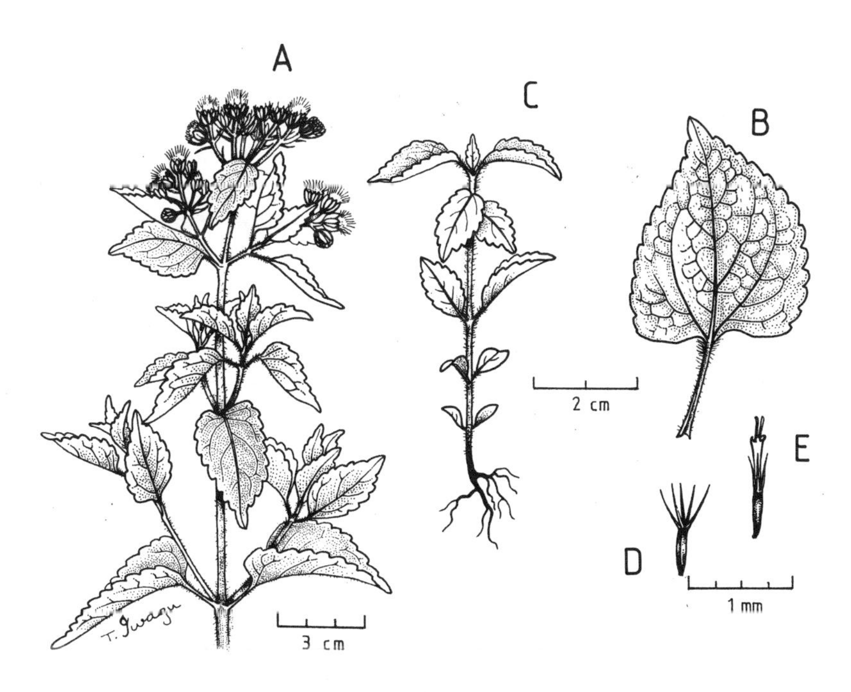

Fig. 13 *Ageratum conyzoides* L.
A Habit diagram, **B** Leaf, **C** Small plant, **D** Seed, **E** Floret.

Medicinal use. In Masangko, the leaves are boiled to make a soup which is then drunk by pregnant women experiencing labour pains. In Nasingalatu, the leaves are heated over a fire and the extracted juice is massaged on the body to relieve headache and side pains.

In other parts of the country, the plant has many uses: in East New Britain and Northern Province, the crushed leaves are mixed with water and drunk to treat diarrhoea; also in East New Britain, fresh leaves are squeezed onto sore eyes, the juice is drunk to arrest vomiting and the crushed leaves are rubbed onto the forehead to soothe a headache (the latter also on Manus Island).

The plant is used to treat diarrhoea in India and Mexico (Watt et al 1962), Malaysia (Burkill 1966) and New Caledonia (Rageau 1973). In the West Indies, the leaves are used as a purgative and to reduce fever (Ayensu 1981).

Chemistry. Alkaloids are not present in the leaves or stem (Hartley, 1973; Holdsworth, 1983). This species contains the chromenes, precocenes 1 and 2. These chromenes induce sterility in insects (Fagoonee, 1981). The essential oil contains ageratochromene dimer (Kasturi, 1973). Conyzorigum, a chromone, is also present (Adesogan, 1978). Nair (1977) found stigmast-7-en-3-ol, quercetin, kaemferol, fumaric and caffeic acids were present. Quercetin has anti-inflammatory and anti-spasmodic properties; kaemferol has anti-inflammatory and diuretic properties and caffeic acid has anti-tumour properties (Duke, 1985).

Local name	Village name
Sikong	Masangko
Noningkepek	Nasingalatu

<u>*Bidens pilosa* L.</u> WEI-M-09
(L. bi – two; dens – tooth)

Description. Herb, 0.5–1 m tall. Leaves opposite (the upper ones may be alternate); lower leaves with 2–3 pairs of pinnae, upper leaves trifoliate. Leaves finely toothed. Heads many flowered, yellow. Fruit an angled achene.

Ecology. Sporadic; found in sparce *Imperata* stands.

Distribution. No information.

Medicinal use. In Manki, the flower of this herb is pressed into the centre of a boil to extract the pus, then the leaf is used as a plaster over the boil.

This plant is used in a similar way in the Southern Highlands. Similarly, in the Philippines, the leaves are used to treat boils (Perry 1980) and in the West Indies, the leaves are used as plasters for sores (Ayensu 1981). In Brazil, the leaves are applied to ulcers (Chopra 1956).

Chemistry. No information could be found.

Local name	Village name
Kwipo	Manki

Blumea arfakiana Mart.

(after K. L. Blume, 1796–1862; Director of the Rijksherbarium, Leiden; type from the Arfak Mountains, New Guinea)

Description. Erect herb or undershrub, 1–2 m in height. Stems woody at base, branched, erect, pubescent. Leaves obovate to elliptic, finely toothed. Flowers arranged in terminal panicles. Corolla yellow, tubular. Achenes with white pappus.

Ecology. Found in rain forests, along river and stream banks, in old gardens and in cleared and settled areas from sea level to 1600 m. Along gravelly and rocky stream banks and creeks, fallow gardens, open places in forests. Flowers March to October.

Distribution. Endemic to New Guinea; occurs in the Moluccas, Aru Islands, New Guinea, Bismarck Archipelago, Solomon Islands, Palau Islands, Somoa Islands.

Medicinal use. In Aseki, the leaves are eaten with traditional salt to treat anaemia; this treatment may continue for several days. In New Britain, the leaves and roots have been reported to treat stomach ache (Futscher 1959).

Chemistry. No alkaloids are present in the bark or leaves (Hartley, 1973). A monoterpene is present in *B. wightiana* oil (Bohlmann, 1979).

Local name	Village name
Kumpeka	Aseki

Erechtites valerianifolia (Wolf.) DC. (Fig. 14)

Description. Erect or decumbent herb, 45–140 cm tall. Leaves petiolate, oblong, pinnately lobed. Lobes acutely dentate. Flowers in heads on long peduncles. Corolla of marginal flowers and disc flowers. Achenes with white hairs.

Ecology. A weed of roadsides and cultivation, rarely troublesome in pasture. Also occurs in landslips and other naturally disturbed habitats. Grows on abandoned native peanut and sweet potato gardens, in dry open roadsides, on grassy hills, riverbanks, creekbeds, on recent volcanic deposits, along swamps, in open places in rainforests and secondary forest, in bamboo forests, ravines, on loamy, sandy black, peaty soils, from 100–2600 m

Distribution. Native to Brazil, introduced to Malesia. Naturalized in Philippines, Sri Lanka, Malay Peninsula, Indonesia, Taiwan,

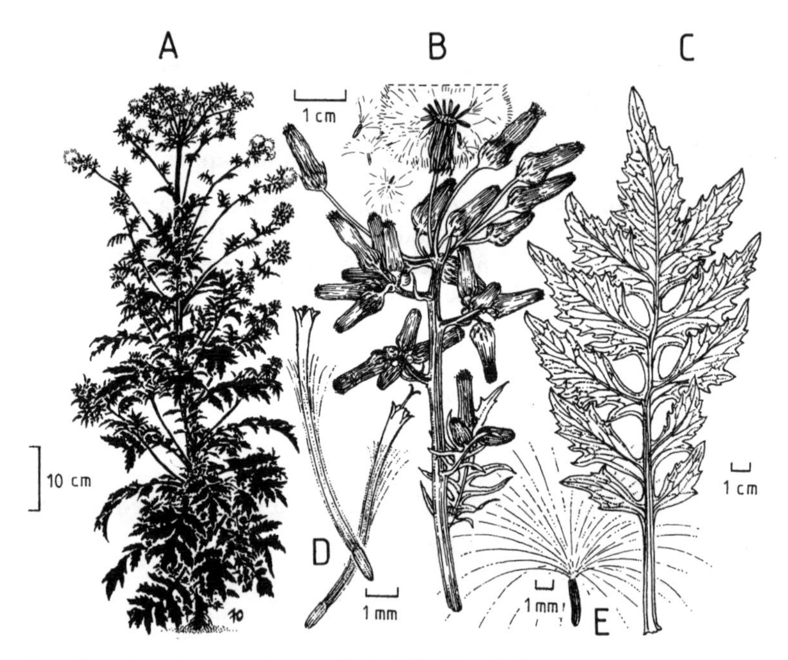

Fig. 14 *Erechtites valerianifolia*(Wolf.) DC.
A Habit diagram, **B** Inflorescence, **C** Leaf, **D** Seed, **E** Floret.

New Guinea, Japan, North Australia, Fiji Islands, Solomon Islands, Hawaiian Islands, Tonga and Somoa. Widespread in New Guinea.

Medicinal use. In Buang,the leaves are repeatedly rubbed on skin infections.

In the southern Highlands, the juice extracted from the leaves and stem is applied to sores and scabies.

Chemistry. No alkaloids were found in this species (Hartley, 1973). The related species *E. quadridentata* contains the alkaloid retrorsine N-oxide (Dictionary of Organic Compounds D-05057).

Local name	**Village name**
Baru sake	Buang

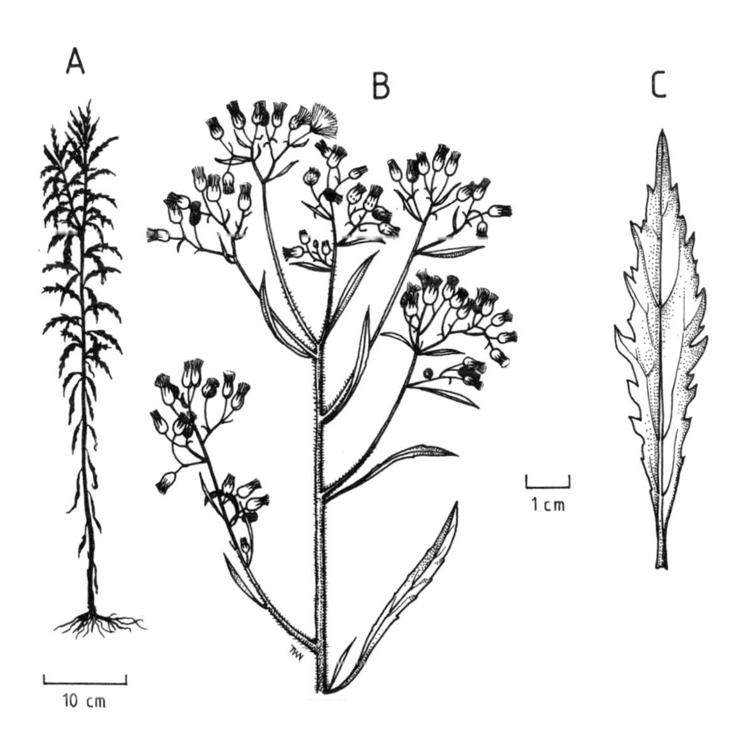

Fig. 15 *Erigeron sumatrensis* Retz.
A Habit diagram **B** Inflorescence arrangement, **C** Leaf.

Erigeron sumatrensis Retz. (Fig. 15)
E. linifolius Willd.
(Gr. eri – early; geron – grey)

Description. Erect, annual herb, 25–160 cm. Stem grooved with long patent hairs. Leaves linear. Flower heads numerous, in dense panicles. Corolla campanulate, pale yellow to yellow-ochre to creamy or white. Marginal flowers pink; disc flowers yellow. Achenes light brown with long pappus hairs, later turning rusty.

Ecology. A weed of roadsides, cultivation and plantations, (particularly in young, unshaded coffee or tea), fallow, neglected gardens, meadows, open places, savannah, creekbeds, grasslands in riverplains, tree-fern grasslands, steep clay banks, edge of mossy forest, roadsides, rocky soil, from sea level to 3350 m.

Distribution. Native to Malesia; occurs in Malay Archepelago, Malay Peninsula, Philippines, Thailand, Fiji, Tonga, New Guinea.

Medicinal use. In Buang, the young leaves are rubbed on the skin to treat the skin fungal infection, grille. A related species, *E. canadense*, is used as a treatment for eczema in South Africa (Watt 1962).

Chemistry. No information could be found.

Local name	Village name
Guhehane	Buang

Microglossa pyrifolia (Lamk.) Kuntze (Fig. 16)
WEI-Bun-75

Description. Shrub, tall herb or upright vine, to 4 m tall; branched. Leaves petiolate, blade ovate-elliptic. Flowers in branched corymbs. Corolla 3 mm long, (pale) yellow-greenish-white. Achenes with a long silky pappus.

Ecology. Found climbing or scrambling on tree tops, along roadsides and riversides, on deforested slopes, in low or young secondary forest, in waste places, in montane mixed forest, in (light) rain forests (edges and clearings), along water races, in gulleys above water; from 240–2400 m.

Fig. 16 *Microglossa pyrifolia* (Lamk.) Kuntze

Distribution. Occurs in India, China, Burma, Thailand, Vietnam, Laos, Malay Archipelago, Malay Peninsula, Philippines, Formosa, New Guinea, tropical Africa, Madagascar, Comores.

Medicinal use. In Buang, the steam from leaves heated over a fire is blown onto a spear wound or a sore eye.

In Northern Province, the juice from crushed leaves is applied to an ulcer which is then covered with the crushed leaves.

Chemistry. No alkaloids found in the stem or leaves (Hartley, 1973).

Local name	Village name
Gogo	Buang

<u>*Mikania scandens* Willd.</u> (Fig. 17)
(after J. C. Mikan, d. 1844, Chech. Botanist)

Description. Climber with a thin stem. Leaves cordate-deltoid, petioled. Inflorescence terminal and axillary, paniculate, cylindrical heads. Achene black with whitish pappus, 2.5–3 mm long.

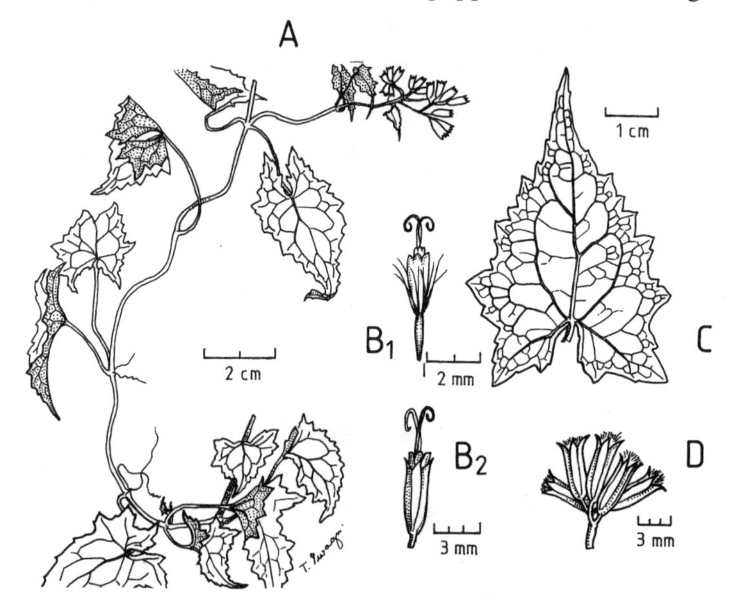

Fig. 17 *Mikania scandens* Willd.
A View of the plant, **$B_{1/2}$** Florets, **C** Leaf, **D** Inflorescence.

Ecology. No information.

Distribution. Occurs in Tropical America; introduced to the Philippines, Mariana Islands, Fiji Islands, Solomon Islands, New Britain, New Guinea.

Medicinal use. In Keregia, the leaves are squeezed and applied to cuts, axe or knife wounds.

In Simbu Province, the stem is squeezed, mixed with ginger and baked in a bamboo stem and eaten with greens to give relief to colds, headache or stomach ache. In the Eastern Highlands, the stem sap is used directly for the same ailments.

In Somoa, also, the leaves are used to treat coughs (Uhe 1974) and in Malaysia, a related species, *M. amara* is used to lower fever (Burkill 1966).

Chemistry. No alkaloids are present in the stem or the leaves (Hartley, 1973). Twenty seven terpenoid constituents, (-)-kaur-16-en-19-oic acid, (-)-kauren-16-ol and taraxasterol were isolated from the essential oil of *M. micrantha* (whole plant) (Nicollier, 1981). Hertz (1975) reports the presence of the sesquiterpene mikanokryptin.

Local name	Village name
Hawec muc	Keregia

Wedelia biflora (L.) DC. (Fig. 18)

(after G.W. Wedel, 1645–1721, professor)

WEI-Z-03, WEI-Bu-94

Description. Herb, or sub-shrub, 1.5–2.5 m tall, more or less climbing, dull green. Leaves opposite, ovate, toothed. Flower heads 18–38 mm across, in irregular panicles. Ray flowers female, bright yellow; disc flowers tubular, bisexual and darker in colour.

Ecology. Common everywhere, in secondary forest, disturbed areas.

Distribution. Occurs India to Polynesia.

Medicinal use. In Buso, the bark from the root is scraped and drunk in cold water to treat diarrhoea. Similarly, in the Finschhafen area, the inner stem is burned over a fire and the resultant ash is mixed with water and drunk to treat diarrhoea and dysentery. Only roasted foods may be eaten during the course of treatment.

This plant is used extensively in Papua New Guinea for a number of ailments. In Northern Province, Buka Island (North Solomons Province), and the Sepik, the leaves are crushed, mixed with water and used to treat bad coughs. In New Britain, the stem is crushed in water to treat diarrhoea; the leaves are used to relieve dysentery and stomach ache in New Ireland, Madang, Manus Island, and Bougainville. Malarial fever is treated with crushed leaves in sea water in New Britain and Buka Island; crushed leaves are used to relieve headaches on Manus Island and Buka Island, where it is also chewed to soothe toothache. In Central Province and Manus Island, the leaf sap is used to clot the blood from a heavily bleeding cut and to aid in scab formation. In New Ireland and Madang, cuts and sores are treated with the juice from the soft stem and leaves. In Somoa, the leaves are used to relieve stomach ache (Uhe 1974); in Thailand, the leaves are used in malaria treatment (Burkill 1966) and in Tonga, the plant is used to treat wounds (Weiner 1971).

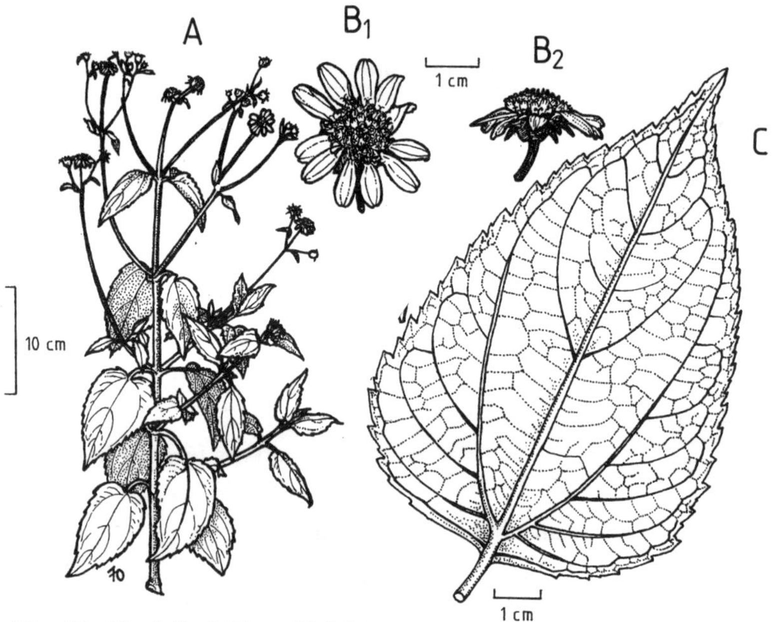

Fig. 18 *Wedelia biflora* (L.) DC.
A Habit diagram, $\mathbf{B_{1/2}}$ Flower, **C** Leaf.

Chemistry. No alkaloids are present (Hartley, 1973; Holdsworth, 1983). Some species of *Wedelia* contain lactones (Bohlmann, 1980, 1981; Govidachari, 1956, 1957). A potential anti-tumour compound, wedeloside was isolated from *W. asperrima* (Oelrichs, 1980).

Local names	Village names
no record	Buso
Qasauc	Keregia
Igiac	Nasingalatu
Mangali	Zafiruo

CONVOLVULACEAE

Merremia peltata (L.) Merr. (Fig. 19)
Ipomoea peltata Choisy WEI-Sq-08,WEI-S-05

Description. A large climber or twiner (liana), to 30 m high, covering whole trees, occasionally procumbent. The stems form a large subterraneous tuber with milky juice. Leaves peltate, petioled. Inflorescence to 40 cm long, widely corymbose. Flowers several to many, large sepals (18–25 mm). Corolla white or yellow, large, broadly funnel shaped, 4.5–6 cm long. Fruit a capsule, opening by 4 valves, splitting longitudinally. Seeds yellowish, dark brown, densely tomentose and long villose. The tubers are edible.

Ecology. Common; found in the edges of primary and secondary forests, clearings, thickets, from sea level to ca. 700 m.

Distribution. Occurs in Madagascar, Mascarenes, Seychelles, Malay Pennisula, Malay Archipelago, Philippines, New Guinea, N & E Australia, Polynesia.

Medicinal use. In the Finschhafen area (Suquang), the stem sap is applied to knife and axe wounds and in Sosoningko, it is drunk to soothe a cough or a headache.

A similar treatment is used on Manus Island, where a piece of the stem is put in the mouth and a bubble of sap is blown onto a new cut. Also on Manus Island, new leaves are placed on sores. On Karkar Island, the stem sap is used on boils.

Fig. 19 *Merremia peltata* (L.) Merr.

The plant has similar uses in other countries: it is applied to wounds and swellings in Indonesia (Soepardi 1967), to wounds and sore breasts in Malaysia (Burkill 1966) and to bruises in Somoa (Uhe 1974).

Chemistry. Alkaloids are not present in the leaves or the stem (Hartley, 1973; Holdsworth, 1983). Some species do contain alkaloids.

Local names	Village names
Wafigu	Suquang
Bororo	Sosoningko

CRASSULACEAE

Kalanchoe pinnata Lamk. Pers. WEI-F-19
Bryophyllum calycinum Salisb.
B. pinnatum Lamk. Kurz

Description. A robust, unbranched herb or sub-shrub, 0.3–2 m high. Perennial; succulent. Leaves opposite with crenate margins. Flowers large, arranged in cymes, calyx purple, corolla green in the lower half and red in the upper half.

Ecology. Plants of sandy, stoney and sunny localities, always dry and never far from human habitation. Locally often gregarious. Often cultivated as a garden plant and for its medicinal properties. In PNG it is cultivated from sea level to at least 2000 m.

Distribution. Possibly originated in tropical Africa but at present pantropical. Introduced to Malaysia long ago.

Medicinal use. In Fondengko, the leaves are heated in an open fire and are chewed to relieve toothache. In Sosoningko, the leaves are squeezed onto boils which are then wrapped with leaves of *Gnetum gnemon*. The treatment is repeated daily until the boil bursts and heals.

On Karkar Island, the leaves are heated and used as a poultice for sores and swellings.

Similarly, in the West Indies, the bruised leaves are used as a dressing for boils, bruises, ulcers and insect bites (Ayensu 1981).

Chemistry. Alkaloids are not present (Holdsworth, 1983).

Local names	Village names
Qinzing	Fondengko
Aika	Mabsiga
Kalucluc	Nasingalatu

CYCADACEAE

Cycas circinalis L. WEI-S-02
C. rumphii
Tree Fern

Description. Tree, stem 3–10 m high. Leaves 1–2 m long, pinnate. Dioecious; male flowers united in a cone.

Ecology. No information.

Distribution. No information.

Medicinal use. In Sosoningko, the seed is opened and scraped and applied to tropical ulcers for a very effective treatment.

In three different areas investigated in Central Province, and a village in Sepik, the same practice occurs. In another area in Central Province, the leaves are boiled and drunk to soothe a cough. In Burma, Singapore and Malaysia, the plant is also used to treat sores and ulcers (Burkill 1966).

Chemistry. The seeds contain alkaloids; the leaves do not (Hartley, 1973). The nuts also contain a toxic (carcinogenic) constituent, (methyl-ONN-azoxy) methanol (Dictionary of Organic Compounds, M-00887). A flavone (2,3-dihydrohinokiflavone) has been isolated from the leaves of *Cycas* species (Bechmann, 1971).

Local name	Village name
Tupeh	Sosoningko

DAPHNIPHYLLACEAE

Daphniphyllum gracile Gage WEI-Wau-179

(Gr. daphne – laurel; phyllon – leaf)

Description. Glabrous shrub 1.5–5(–8) m tall, or tree to 12(–23) m tall, much branched. Leaves usually crowded at the ends of the branches. Petioles long, red coloured, the leaves leathery. The inflorescence in unisexual racemes. Fruit globular, fleshy.

Ecology. Common, from (850–)1000 to 3500 m in primary and secondary lower mountain to subalpine rainforest, on margins of subalpine grassland and in exposed summit shrubberies.

Distribution. Common in the mountains throughout Papuasia, in New Britain and New Ireland.

Medicinal use. In Aseki, the leaves of this plant species are chewed and swallowed with traditional salt to treat hookworm.

Similarly, in Japan, a related species, *D. macropodium* is used as a vermifuge (Perry 1980).

Chemistry. Six alkaloids have been isolated (Yamamura, 1977). Some of the *Daphniphyllum* alkaloids act on the central nervous system (Manske, 1975).

Local name	Village name
Demaiyo	Aseki

DILLENIACEAE

Tetracera nordtiana F. Mueller WEI-F-02

Description. A shrub or large climber, to 10 m high. Leaves elliptic-lanceolate and petioled. Inflorescence terminal, 15–50 flowers. Corolla with 3 petals. (This is a variable species with 6 varieties).

Ecology. Occurs in Australia (Queensland), in Malaysia: New Guinea, Louisiades, SE Sulawesi and Moluccas.

Distribution. No information.

Medicinal use. In Bolinbaneng, the whole plant is wrapped in banana leaves and heated over a hot fire, then the liquid is squeezed from the cooled plant and drunk to treat fever or influenza. In Fondengko, also in the Finschhafen area, the stem sap of *Tetracera* sp. is drunk to treat body pains, headache and pneumonia.

Chemistry. Alkaloids are not present in the leaves (Hartley, 1973; Holdsworth, 1983). This genus contains the triterpenes, lupeol, betulin and betulinic acid and B-sitosterol (Dan, 1980). Lupeol, betulin and betulinic acids are claimed to have anti-tumour properties. B-sitosterol is claimed to be antihypercholesterolemic, anti-prostatic, anti-prostatadenomic and estrogenic (Duke, 1985).

Local names	Village names
Quinzin	Bolinbaneng
Kauwang	Fondengko

Dillenia castanefolia (Miq.) Diels

(after the German botanist, J. J. Dillenius 1684–1747)

Description. Tree up to 20 m high. Leaves oblong to elliptic-oblong, petioled. Flowers in racemes, 1–6 flowered. Petals a deep lemon yellow. Fruits are dehiscent with enlarged red sepals.

Ecology. Found in primary and secondary forests, usually on riversides.

Distribution. Occurs in Malaysia: New Guinea and islands in the Geelvink Bay.

Medicinal use. The bark is chewed with traditional salt and spit into a bamboo container which is then inhaled or allowed to drip into the nose to treat a cold, chest pains or a low grade fever.

A related species, *D. indica*, is used to treat fever in India (Chopra 1956) and Burma (Perry 1980) and the juice of *D. philippenensis* is used to treat pains in the chest. (Perry 1980).

Chemistry. The genus has been found to contain the triterpenes lupeol, betulin and betulinic acid and B-sitosterol (Dan, 1980).

Local name	Village name
Wekuwa	Aseki

ELAEOCARPACEAE

Elaeocarpus sphaericus (Gaertn.) K. Sch. (Fig. 20)

Description. Large tree, up to 40 m tall. Leaves usually scattered, petioled. Usually with a thin widespreading buttress. Inflorescences borne on twigs behind the leaves. Fruit globose, bright blue or purple.

Ecology. A common secondary species in disturbed vegetation, but it can persist to become a canopy or emergent species. Found in lowland and lower montane rainforests throughout New Guinea.

Distribution. Throughout lowland Papuasia. From India through Malesia to Australia and Fiji.

Medicinal use. In Bolinbaneng, about 100 ml of the sap is collected and diluted with water; this is drunk 2 or 3 times to cure stomach ache or pains in the chest or shoulders.

The fruit of a related species *E. grandiflora,* is used as a mild diruetic in Indonesia (Steenis-Kruseman 1953).

Chemistry. The leaves and the bark contain 10-indolizidine alkaloids (Johns, 1971).

Local name	Village name
Qozari	Bolinbaneng

ERICACEAE

Rhododendron macgregoriae L. (Fig. 21) WEI-Wau-180

(note: there are three varieties; assuming var. *macgregoriae*)
(Gr. a rose coloured tree)

Description. Shrub or treelet, 0.5–5 m high. Leaves leathery, petioled. Flowers in umbels of 8–15 flowers. Corolla tubular, light yellow to dark orange or yellow at the tube and reddish-orange at the lobes. Stamens 10. Fruit a cylindric capsule, slightly curved.

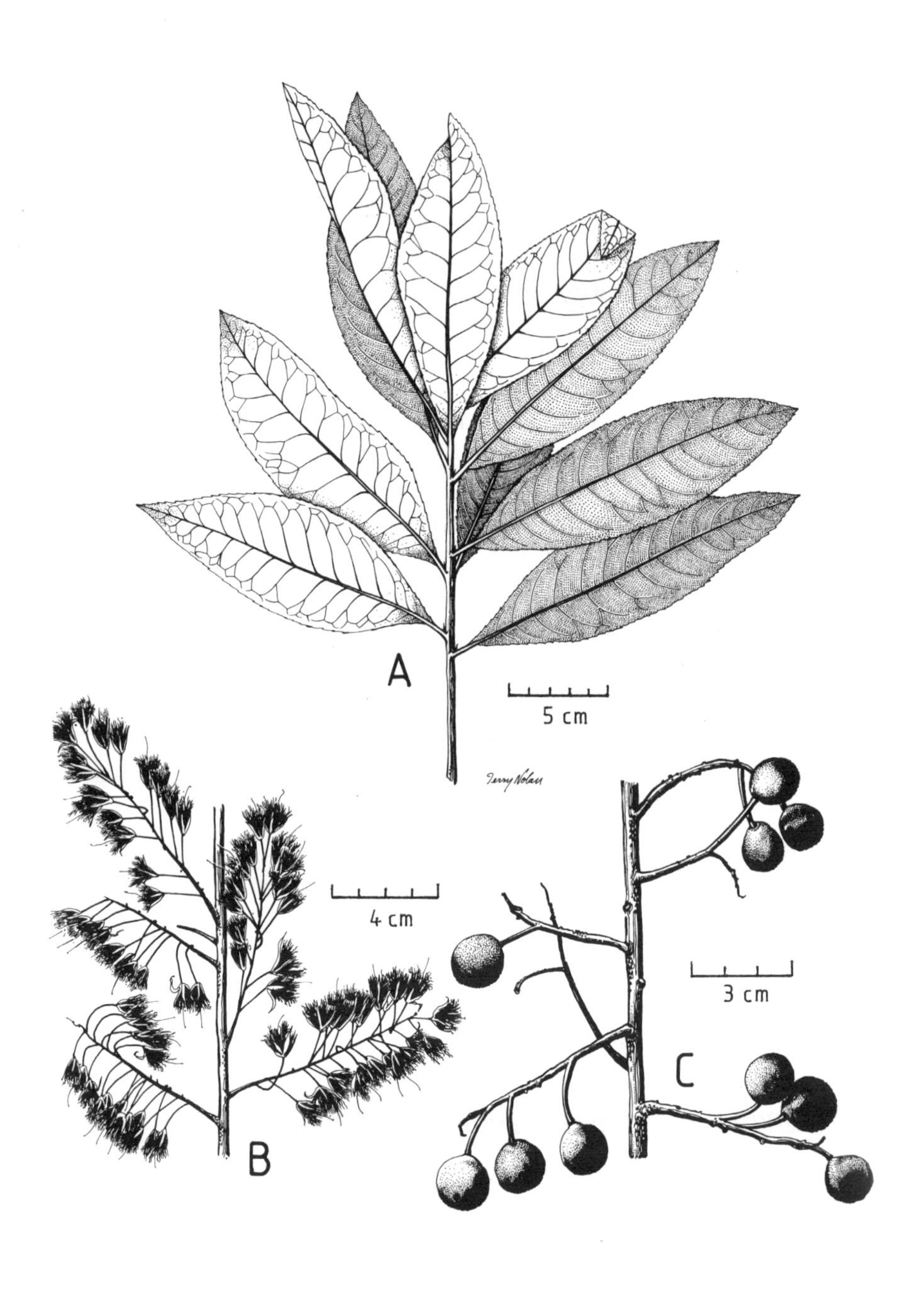

Fig. 20 *Elaeocarpus sphaericus* (Gaertn.) K. Sch.
A Branch, **B** Inflorescence, **C** Fruit arrangement.

Ecology. Fairly common and locally abundant. Found in primary mossy forest, secondary bushes and moderately dry sunny situations, among trees on mountain slopes, in thin canopy shade, also along creeks, on precipitous banks of streams, scattered over open grassland, on landslides and formerly cultivated land, invading artificial clearings (airstrips), on sandy or peaty soil, stony clay and limestone ridges at (120–)500–3000(–3350) m.

Distribution. Occurs only in New Guinea.

Medicinal use. In Aseki, the leaves are crushed and mixed with a few drops of water, then applied to tropical ulcers. The sore heals quickly after some initial irritation.

In Simbu, a similar treatment is used: the sap from the heated young leaves and flowers is squeezed onto ulcers and sores. Reported to be poisonous to mules (Edie Creek).

Chemistry. No alkaloids are present in the leaves or the bark (Hartley, 1973). Triterpenes and flavones are present in some species (Arthur 1955, 1960, 1961). *R. dauricum* is a folk medicine

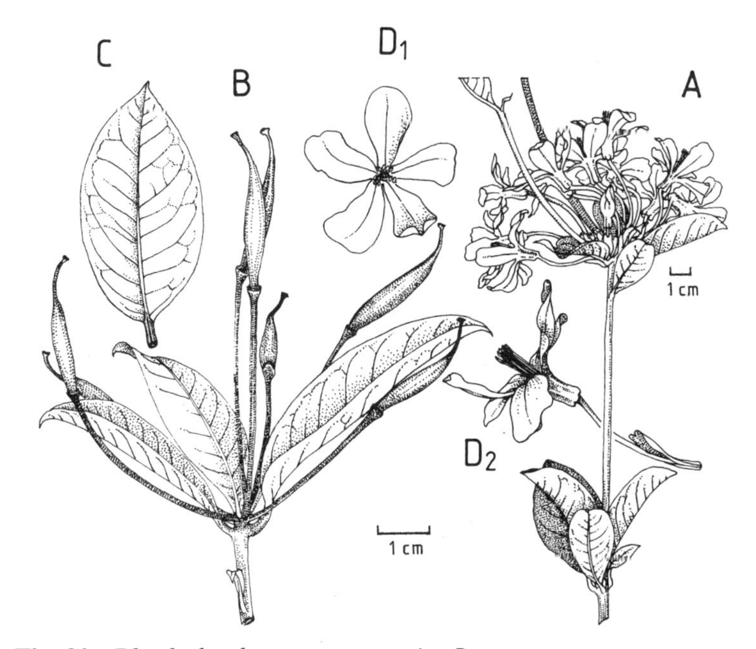

Fig. 21 *Rhododendron macgregoriae* L.
A Habit diagram, **B** Fruit, **C** Leaf, $\mathbf{D_{1/2}}$ Flower

used in the treatment of chronic bronchitis. A flavanone expectorant ingredient farrerol has been isolated (Xiao, 1981).

Local name	Village name
Kanama	Aseki

Vaccinium keysseri Schlechter
(there are two varieties: assuming var. keysseri)
(L. whortleberry)

Description. Erect stiff shrub or straggling tree up to 6 m. Flowers in racemes in upper leaf axils, 10–15 flowered. Corolla campanulate, fleshy pink. Fruit a berry, blackish at maturity.

Ecology. Mostly terrestrial, rarely epiphytic, locally common in montane *Podocarpus* forest, in edges of subalpine (mossy) forest and summit growth (2300–)3000–3600(–3800) m.

Distribution. Occurs only in New Guinea.

Medicinal use. In Aseki, the leaves are crushed and eaten with traditional salt to treat internal sores.

Chemistry. No references to this species could be found. Other species are known to contain anthocyanins. *V. myrtillus* contains an alkaloid (Glosse, 1978).

Local name	Village name
Utitava	Aseki

ERYTHROXYLACEAE

Erythroxylum ecarinatum Burck.

Description. Tree, 7–37 m tall. Crown small, not widely spreading. Leaves narrowly elliptic, abundant twig ends. Flowers in clusters. Petals white, yellow to greenish yellow and cream. Fruit a drupe changing from yellow to orange to red.

Ecology. A common tree which occurs scattered from the lowlands to 2000 m in rainforests on slope and mountain ridges, on rocky and clayey soils.

Distribution. Occurs in NE Australia, Melanesia and east Malaysia; Sulawesi, Moluccas and New Guinea.

Medicinal use. In Aseki, the leaves are chewed with traditional salt and the sap swallowed to treat an upset stomach and to prevent further vomiting.

In Sulawesi, it is used as a medicine. A related species *E. monogynum,* is used in India for stomach upsets and fever (Chopra 1956).

Chemistry. The leaves contain alkaloids (Hartley, 1973), but they have not been identified. *E. coca* contains the alkaloid cocaine.

Local name	Village name
Taa	Aseki

EUPHORBIACEAE

Acalypha insulana Muell. Arg. WEI-Bua-56
A. hellwigii Warb.
(Gr. a – without; kalyphos – cover)

Description. Shrub or tree. Often gregarious, to 3 m high. Short petioled, coriaceous leaves, sharply serrate margins. Lax flowered female spikes, up to 25 cm in length.

Ecology. Found in primary and secondary rainforest, in open savannah, on river banks, in creek beds, grassland, kunai (*Imperata*) and marine strand vegetation.

Distribution. An endemic species, widespread in New Guinea.

Medicinal use. In Buang, a quartz pebble is heated over a fire and laid in the leaves, then hot water is poured over the hot stone and the whole thing is placed on the sore. In the coastal village of Buso, *Acalypha* sp. is brewed in water and the resulting bright red solution is used to bathe a sick person whose illness cannot be explained and so is attributed to possession by spirits.

A related species *A. novoguineensis* is used on the Solomon Islands: the leaves are rubbed on wounds and the steam from boiled leaves is applied to sore eyes (Perry 1980). In Malaysia, the roots of *A. hispida* are used as a poultice for leprosy and the leaves are used for wounds and ulcers (Perry 1980).

Chemistry. Alkaloids are not present in the leaves or the bark (Hartley, 1973). A cyanogenis glucoside acalyphin has been isolated from *A. indica* (Nahrstedt, 1982).

Local names	Village names
Bluwa	Buang
Anasuwe	Buso

Breynia cernua (Poir.) Muell. Arg.

Description. Small tree.

Ecology. Found in regrowth.

Distribution. No information.

Medicinal use. In two mountain villages in the Finschhafen area, the bark of this tree is scraped and mixed with water and drunk to treat dysentery.

In East New Britain, a poultice of the leaves act as a local anaesthetic and are used to ease intense body pains. Also, the leaves are heated with lime in salt water over a fire, then the mixture is rubbed on a sore or tropical ulcer.

In Malaysia, a related species *B. reclinata* is used to poultice skin diseases and swellings (Burkill 1966).

Chemistry. No information could be found.

Local name	Village name
Ziziling	Zazaquo
	Quaqua

Codiaeum variegatum Bl.

(codebo – Malay name)

Description. Shrub, 1–4 m tall. Leaves alternate, leathery, of variable size and colour. Monoecious; male flowers in racemes on pedicels, female flowers without corolla. Fruit a capsule; seed black and shining.

Ecology. Common in gardens. Widespread in the wild in undisturbed land and primary forest.

Medicinal use. The leaves are used in Nauti, to rub on the abdominal area to case abdominal pain.

This plant is cultivated for a number of medicinal purposes in Papua New Guinea. On Manus island, the root is chewed with betel nut to treat stomach ache and tooth ache. Also, the leaves are boiled and a person with fever is bathed in the cooled green water. In New Ireland and Central Province, the leaf sap is squeezed on to sores. In Central Province, the leaves are bitten to strengthen teeth. Also, snake bite is treated by drinking the leaf sap and rubbing it on the wound after the affected flesh has been cut away with a knife. In the North Solomons Province, near Keita, the crushed root is mixed with volcanic sulphur and the mixture is chewed to induce sterility in women.

In Malaysia, the plant is used as a purgative and to treat sores (Burkill 1966).

Chemistry. Alkaloids are not present in the leaves or the bark (Hartley, 1973).

Local name	Village name
Hamawanga	Nauti

Endospermum moluccanum (Teys. & Binn.)

(Gr. endos – within; sperma – seed)

Description. Tree 8–30 m tall. Leaves peltate, broadly ovate, characteristically crowded towards the end of the branches. Dioecious. (Difficult to distinguish from *E. labios*).

Ecology. The trees are found scattered in primary riverine forest, subject to flooding, up to 750 m (0–1000 m).

Distribution. Occurs in Sulawesi, Moluccas, West New Guinea, Bismarck Archipelago (New Britain) and the Solomon Islands.

Medicinal use. In Nasingalatu, the bark is crushed and heated to treat a dislocated knee or arm; crushed and heated leaves are applied to boils in the armpit.

In Indonesia, a decoction of the wood is used to treat ulcers (Perry 1980). The thin roots are used as an antidote against arrow poison.

Young leaves are boiled and consumed as a vegetable producing a soft purgative effect. The old leaves are strongly laxative. "If a man has become short winded as a result of eating food prepared by a menstruating woman he eats the skin of a mango wrapped in these leaves and he puts them on his throat." (Blackwood, 1935).

Chemistry. No information could be found.

Local name	Village name
Kalisic	Nasingalatu

Euphorbia hirta, L. (Fig. 22)

(after Euphorbos, physician to King Juba, of Mauretania c. 54 BC)

Description. Annual herb, up to 60 cm tall. Leaves opposite.

Ecology. Found on roadsides, as a path and garden weed, in clearings, areas of cultivation, in secondary growth, grasslands and vacant lots. Also occurring in natural habitats at streamsides, behind beaches and in grassland.

Found from low alitudes to ca. 2000 m (in New Guinea). Common.

Distribution. A pantropical weed of South American origin. An early introduction to Malesia.

Medicinal use. In coastal Finschhafen, the root is boiled in water and the solution, along with the leaves, is then rubbed on scabies.

In Central Province, the plant is boiled and the solution drunk by a person who passes blood in their urine.

In the Solomon Islands, the fruit is used to treat diarrhoea in children (Foye 1976). The Chinese use the plant to treat fever, dysentery and skin conditions (Chinese Paramedical Journal 1977). In the Philippines and Indonesia, the plant is used to treat bowel problems (Steenis-Kruseman 1953).

Chemistry. Alkaloids are not present (Hartley, 1973). The plant contains triterpenoids, flavanoids and substances known to have antispasmodic and antihistamine properties (Baslas, 1980).

Local name	Village name
Bubu	Nasingalatu

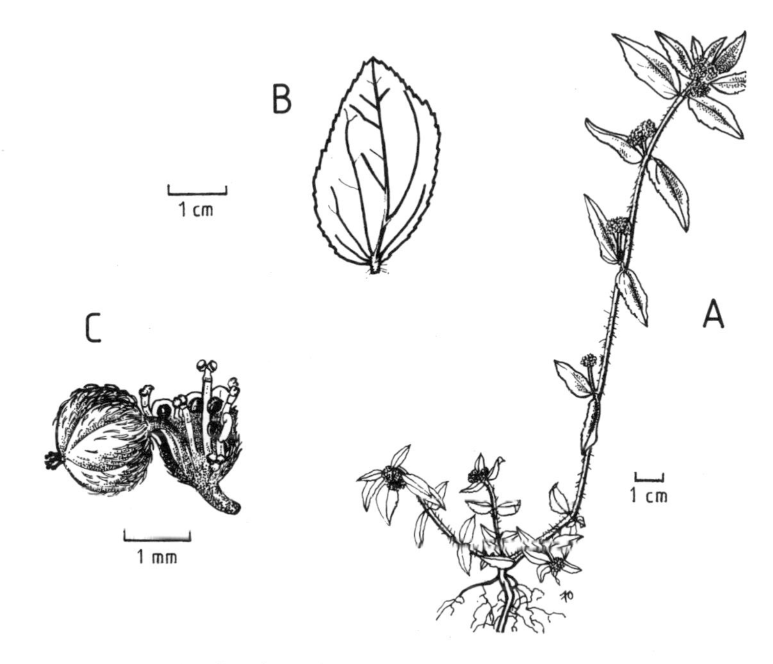

Fig. 22 *Euphorbia hirta* L.
A Habitat of the plant, **B** Leaf, **C** Fruit.

Phyllanthus debilis, Herb. Ham. ex Wall.

(Gr. phullon – leaf; anthos – flower)

Description. An annual weed. Main stem sharply angled. Flowers in unisexual cymules. Fruit a capsule. Similar to *P. amarus*.

Ecology. An introduced annual weed in botanical gardens and airstrips

Distribution. Most probably a native of tropical Asia.

Medicinal use. In two villages in coastal Finschhafen, gonorrhoea and irregular menstrual periods are treated by drinking a cooled solution of the plant boiled in water.

Chemistry. No information could be found. Several species of *Phyllanthus* contain alkaloids. For example, *P. discoides* contains securinine, a central nervous system stimulent similar to strychnine (Manske, 1973).

Local names	Village names
Qinsing	Keregia
Waungwaung	Nasingalatu

Pimelodendron amboinicum Hassk. (Fig. 23)
(Gr. pimelos – fat; dendron – tree)

Description. A large tree, up to 30 m high.

Ecology. Common and widespread in primary and secondary lowland rainforest, swamp forest or mangrove fringes. Occasionally found on coral limestone or scoria or in coastal savannah. From sea level to 800 m.

Distribution. Occurs in Talaud Islands, Sulawesi, Moluccas, Solomon Islands, NE Australia (Queensland).

Medicinal use. In Nasingalatu, a mixture is made of the heated leaf, a leaf of another tree species (*Endospermum moluccanum*) a male prawn, sea water and rain water. This mixture is then boiled, the leaves removed and then drunk to cure a swollen spleen. (Bark may be added if a stronger dose is desired).

In Milne Bay, the leaves are squeezed in water and the solution drunk to treat a cough.

In Malaysia, the leaf is used to clean the mouths of children (Burkill 1966).

Chemistry. Alkaloids are not present in the leaves or the bark (Hartley, 1973).

Local name	Village name
Kalisic	Nasingalatu

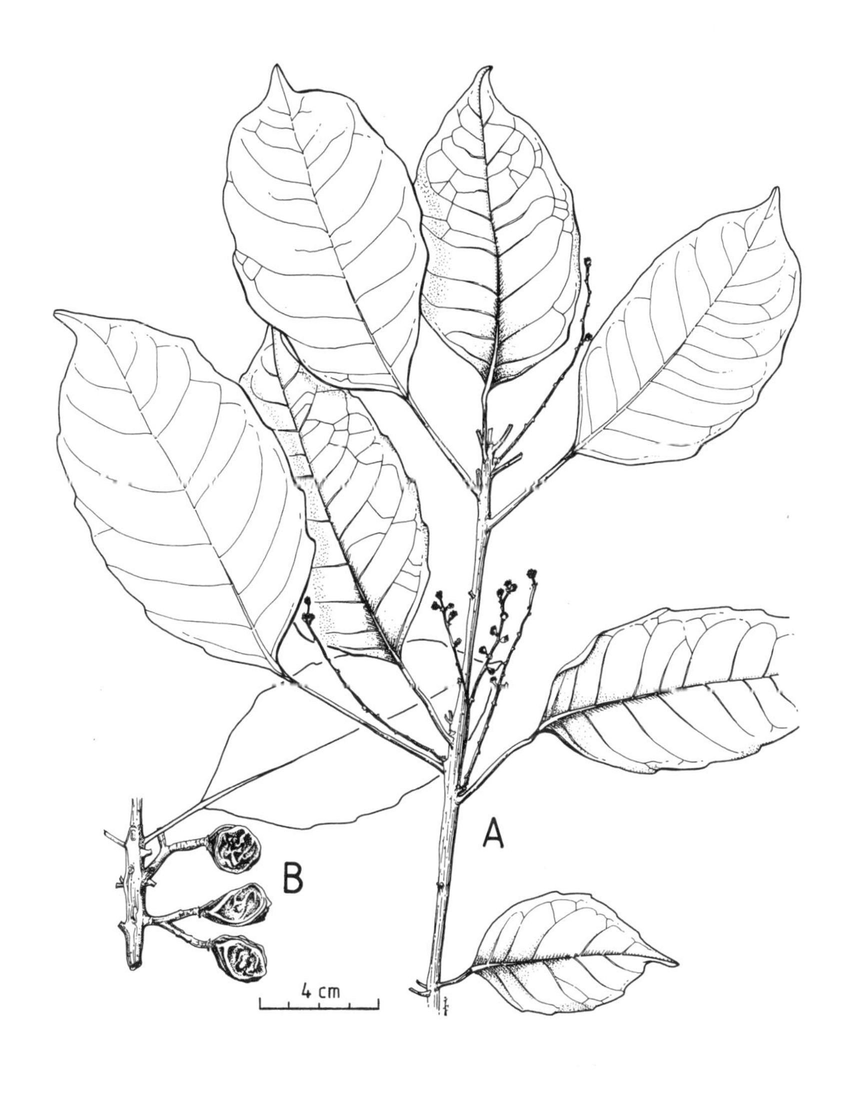

Fig. 23 *Pimelodendron amboinicum* Hassk.
A Habit diagram, **B** Fruit.

FLACOURTIACEAE

Flacourtia rukam Zoll. & Mor.
(after E. Flacourt, 1607–1660, governor of Madagascar;
rukam – Malay name)

Description. Small tree, 5–15(–20) m. Trunk and old branches usually crooked, gnarled and furrowed. Branched near the base. Woody thorns on trunk when young to 4 cm long. Flowers in racemes, unisexual. Corolla greenish-yellow, scentless. Fruit globose, crowned by a ring of small peg-like styles.

Ecology. Found in primary and secondary forest, also in teak forests, often along rivers, from the lowland up to ca. 2100 m in a wild state. It is often cultivated for its edible fruit. Widely distributed but scattered.

Distribution. Occurs both wild and cultivated all over Malaysia, apparently rare in the Moluccas and New Guinea.

Medicinal use. In Mabsiga, the sap from the heated leaves is placed on a boil and thc lcaf is used as a bandage. When the boil bursts underneath, it is cleaned by the dressing.

In New Britain, the leaves are similarly used on wounds, the fruit is ingested to treat diarrhoea and dysentery and the leaf sap is used on eye irritations (Futscher 1959).

Chemistry. Some species contain cyclopentenyl fatty acids (Rehfeldt, 1980).

Local name	Village name
Lumbulum	Mabsiga

FLAGELLARIACEAE

Flagellaria indica L. WEI-YB-07, WEI-Bu-113
(L. flagellum – whip)

Description. A perennial climber, 2–5 m long. Flowers in panicles bearing flowers in short, dense spikes. White, odorous flowers. Fruit a subglobose, smooth, pink drupe, 0.6 mm diameter.

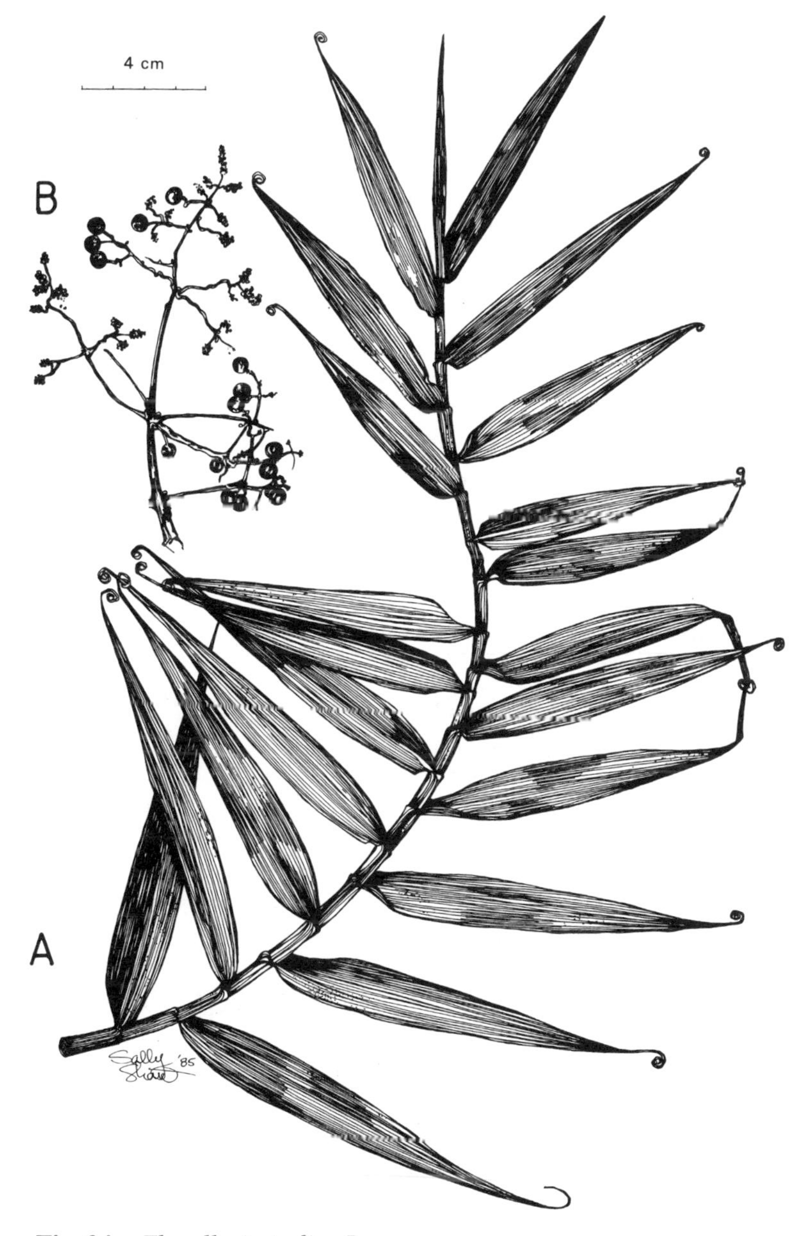

Fig. 24 *Flagellaria indica* L.
A Habit diagram, **B** Fruit.

Ecology. Found in moist (not swampy) forests, along forest borders and the inner margins of mangrove forests.

Distribution. Occurs in tropical Africa, Sri Lanka, tropical SE Asia, Malaysia, Melanesia and Polynesia to North Australia.

Medicinal use. In three villages near Finschhafen, the leaf is heated and the warm sap is blown onto a sore eye. In Yambo, the leaves are crushed and the juice is applied to sores and tropical ulcers.

In the Sepik, the stalk is chopped into small pieces, mixed with water, filtered and then drunk to relieve stomach ache.

In both the Philippines and Malaysia, the flower, leaf and stem are used as a diuretic (Quisumbing 1951; Burkill 1966). It has also been found to be used as a contraceptive for women (Webb 1960).

It is used as a binding and plaiting material, commonly used for sewing sago matting.

Chemistry. Alkaloids are not present (Hartley, 1973). Flavanoids are present.

Local names	Village names
Soangang	Keregia
Mung	Nasingalatu
Suwagin	Suquang
Mingop	Yambo

GOODENIACEAE

Scaevola sericea Vahl. (Fig. 25)
S. taccada (Gaertn.) Roxb. WEI-YB-03, WEI-Bu-98
(L. scaevus – improperly developed)

Description. An erect or spreading shrub, sometimes a small tree (up to 7 m). Leaves spirally arranged, no petiole, a cluster of hairs in the axil. Flowers white, arranged in cymes. Fruit a globular drupe, white when ripe.

Ecology. Restricted to the seashore, usually on open, sandy beaches or coastal cliffs. Common, gregarious. It is an early eure-

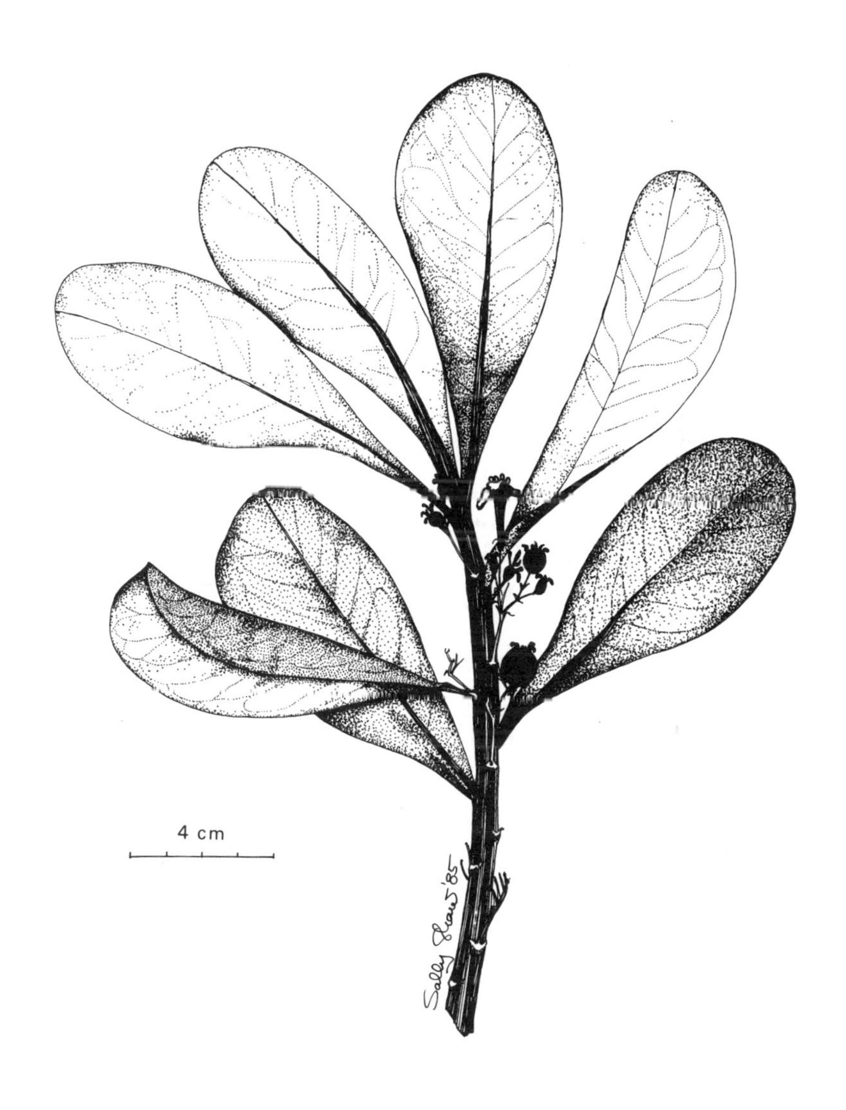

Fig. 25 *Scaevola sericea* Vahl.

daphic pioneer on fresh littoral sands, gravel and rocks. Flowers and fruits the year round.

Distribution. No information.

Medicinal use. In a Finschhafen area village, the young leaves are chewed or boiled into a tea or else the juice from heated leaves is mixed with water to treat a cough. The sap may be applied directly to a sore. In Yambo, the leaf epidermal layer is removed and the rest is chewed to treat malaria.

Similarly, on the islands of Manus, Karkar, Tami and in New Hanover, the leaves are used to treat a cough or cold. In New Ireland and Milne Bay, the plant is used as a form of contraception by women. Other uses include the treatment of eye infection in Milne Bay, an aching ear on Manus Island and asthma and tuberculosis on Karkar Island.

In Indonesia, the plant is used to treat sore eyes (Burkill 1966).

Chemistry. Alkaloids are not present in the leaves, bark or the flowers (Hartley, 1973). The leaves give a positive test for saponins (Simes, 1959). Two glycosidic compounds of unspecified nature have been found (Everist, 1974).

Local names	Village names
Akajok	Mabsiga
Gasoc	Yambo

GRAMINAE

Eleusine indica Gaertn. WEI-Q-16

Description. Grass, stem 30–60 cm tall, only the lower half with leaves, 15–25 cm long. Flowers in 3–7 terminal spikes.

Ecology. No information.

Distribution. No information.

Medicinal use. In Quaqua, the whole plant is uprooted, washed and chewed for the treatment of diarrhoea and dysentery.

In other parts of the country, uses are quite different; in Northern Province, the leaves and stem are boiled and the solution drunk to stop vaginal bleeding. On the coast in Central Province, the whole plant is crushed and filtered through coconut fibre and applied to wounds.

The plant is used in the Philippines to treat sprains and dislocations (Webb et al 1962). In Malaysia, the grass is used in childbirth and also to strap broken limbs (Burkill 1966).

Chemistry. Alkaloids are not present (Hartley, 1973).

Local name	Village name
Iquazi	Quaqua

Paspalum conjugatum Berg. (Fig. 26) WEI-Sq-01
(Gr. paspalos – millet; L. conjugatum – paired)

Description. A perennial grass.

Ecology. An aggressive weed of cultivation. Common to roadsides, rough lawns, pastures and a particular weed of plantations,

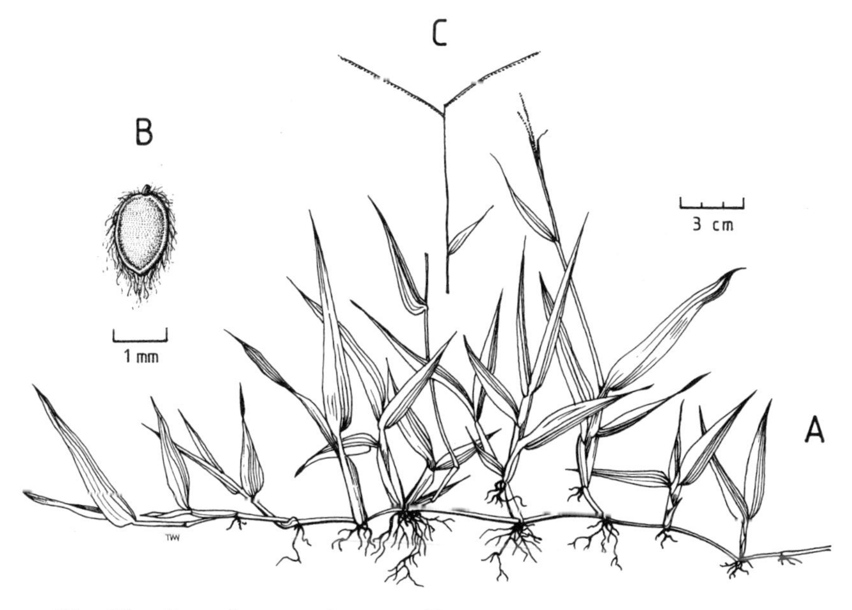

Fig. 26 *Paspalum conjugatum* Berg.
A Habit diagram, **B** Seed, **C** Inflorescence.

where it can become a serious pest, competing with new crops and shade trees. Tolerant to partial shade, from sea level to 2000 m.

Distribution. It is native to tropical America, naturalized in most tropical countries. It is a comparitively recent introduction to many parts of New Guinea.

Medicinal use. In two villages of the Finschhafen area, this grass is used to treat diarrhoea. In Suquang, the solution from squeezed leaves is drunk and applied to sores and in Keregia, a decoction of the leaf tips is drunk. The appropriate dose of this solution is about one teaspoonful for adults and half this for children. In Buang, the juice from the grass is applied to new cuts (and said to sting like iodine solution).

In the Sepik, the flower bud is squeezed onto small sores and scratches. In Madang, new leaves, when heated, are used to check bleeding and the roots and sap have been reported to be used for sore throats and stomach ache on Bougainville Island (Blackwood 1935).

Chemistry. Alkaloids are not present (Hartley, 1973). The methyl ether of lupeol (a widespread pentacyclictriterpene) is present in *P. dilatatum* (Dictionary of Organic Compounds L-00498).

Local names	Village names
Kegbang	Suquang
Kecqang	Keregia
Kuang	Buang

<u>*Pennisetum macrostachyum* Trin.</u> (Fig. 27) WEI-Sq-07
(Gr. penna – feather; seta – bristle)
Cat's Tail

Description. Perennial grass, 3–5 m high. Spikelets small, with long bristles. Spikes 25 – 35 cm long, erect or nodding.

Ecology. Forms clumps in open grassland, at forest edges and along stream banks, from sea level to at least 2000 m. Cultivated in gardens.

Distribution. Occurs in Southern Asia to Australia. Found throughout New Guinea.

Fig. 27 *Pennisetum macrostachyum* Trin.

Medicinal use. In Suquang, the young leaves and stem are squeezed and the juice is applied to a centipede bite to ease the pain. In Buang, the young leaf tips are heated and squeezed into the ear to ease earache.

Chemistry. No information could be found.

Local names	**Village names**
Kosi	Suquang
Weling	Buang

Polytoca macrophylla Benth. (Fig. 28)
(Gr. polys – many; tokos – progeny)

Description. A robust perennial, to 3 m in height. Monoecious, spikelets unisexual, the male flowers above the female on the same spike.

Ecology. Common, found in forest edges, roadsides, streambanks, savannah, from near sea level to 1800 m.

Distribution. Found in New Guinea and neighbouring islands.

Medicinal use. The traditional use of this grass is similar to the preceeding one: in Buang, the leaf shoots are heated and squeezed into a sore ear. The treatment may be repeated when necessary.

In New Ireland, the juice is squeezed from the shoot to treat sores on the tongue of a baby.

Chemistry. No information could be found.

Local name	**Village name**
Bagona	Buang

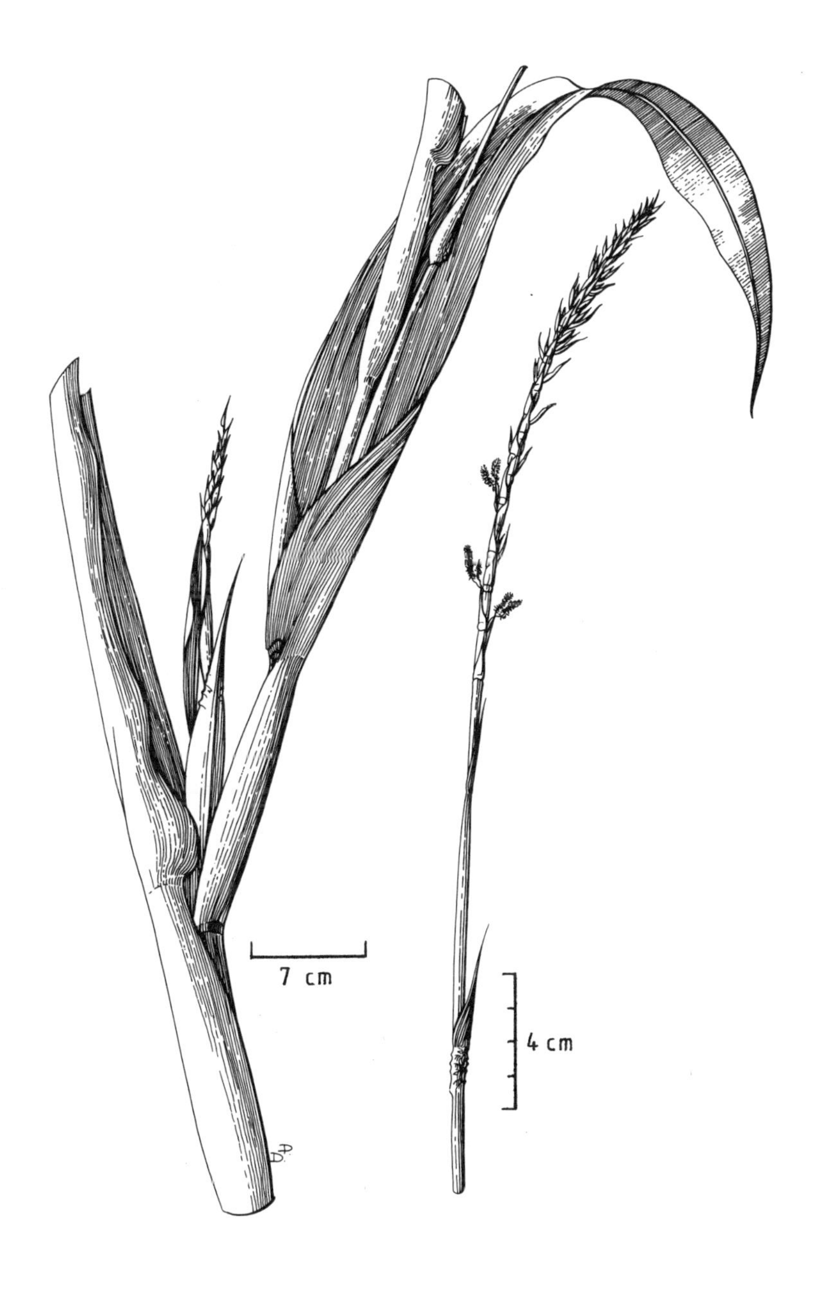

Fig. 28 *Polytoca macrophylla* Benth.

GUTTIFERAE

Calophyllum inophyllum L. (Fig. 29)
(Gr. kalos – beautiful; phyllon – leaf; inos – nerve)

Description. Large tree, 10–20 m tall, bole commonly leaning, often almost horizontal. Leaves opposite, leathery, lateral nerves very fine and parallel. Flowers in axillary racemes, white, 4 sepals and 8 petals. Fruit green, globose, drying pale brown, coarsely wrinkled.

Ecology. Common on beaches immediately above the waterline, often leaning over the sea.

Distribution. No information.

Medicinal use. In Buso, the milky latex from the leaves is diluted with water and the solution is applied to irritated eyes.

All parts of this tree have traditional medicinal uses.

Chemistry. Alkaloids are not present (Hartley, 1973). Phenylcoumarins have been found in the nuts and leaves (Polansky 1957, 1969). Triterpenes are also present (Govindachari, 1967). Calophyllic acid (Polansky, 1957) and calophynic acid are also present (Gautier, 1972).

Local name	Village name
Autawe	Buso

HIMANTANDRACEAE

Galbulimima belgraveana (F. Muell.) Sprague (Fig. 30)
WEI-Wau-182

Description. A large aromatic evergreen tree with a densely compact crown. Up to 35 m tall, sometimes with buttressing. Leaf margin entire. Flowers and fruits occur throughout the year. Monogeneric and monospecific.

Ecology. In the mountains of New Guinea, it is common in the oak forests between 1300 and 2000 m.

Fig. 29 *Calophyllum inophyllum* L.
A Habit diagram, **B** Fruit.

Fig. 30 *Galbulimima belgraveana* (F. Muell.) Sprague.
A Habit diagram, **B** Fruit, **C** Inflorescence arrangement.

Distribution. From the Moluccas to Papuasia and Northeast Australia. It is often a canopy tree found at an altitude of 1200–2700 m, especially in *Nothofagus* forests (can be lower).

Medicinal use. In Aseki, the bark of this tree is chewed and spit into a bamboo container with traditional salt. This mixture is swallowed to relieve abdominal and other body pains. The powdered bark when mixed with tobacco (*Nicotiana tabacum*) and ginger (*Zingiber officinale*) is rubbed into the hair to get rid of head lice.

The bark is rich in alkaloids and is chewed by some in the highlands to induce hallucinations. A variety of alkaloids and other substances have been detected in the bark. The bark and fruit are used by the highlanders to bring about dreams. Chewing of the bark may induce a destructive frenzy followed by sleep with dreams of a hallucinatory nature (Hamilton 1960).

Chemistry. Twenty eight alkaloids have been isolated from this species (Ritchie, 1967). The major alkaloids are himbacine, which has anti-spasmodic properties, and himandridine (Duke, 1985). Lignins are also present.

Local name	Village name
Waga	Aseki

ICACINACEAE

Rhyticaryum longifolium K. Sch. & Laut. (Fig. 31)

WEI-Wau-183

Description. A sprawling shrub or small tree, 1–5 m tall. Leaves spirally arranged, entire. The flowers are yellow or greenish, in male and female spikes. The fruit is a red-orange drupe.

Ecology. A substage of primary, sometimes also found in secondary forest from lowland up to 1800(–2500) m, scattered.

Distribution. Occurs in Northeast Australia, Melanesia, in Malesia: New Guinea.

Medicinal use. The sap is extracted then stored in a bamboo container and sniffed to clear a blocked nose.

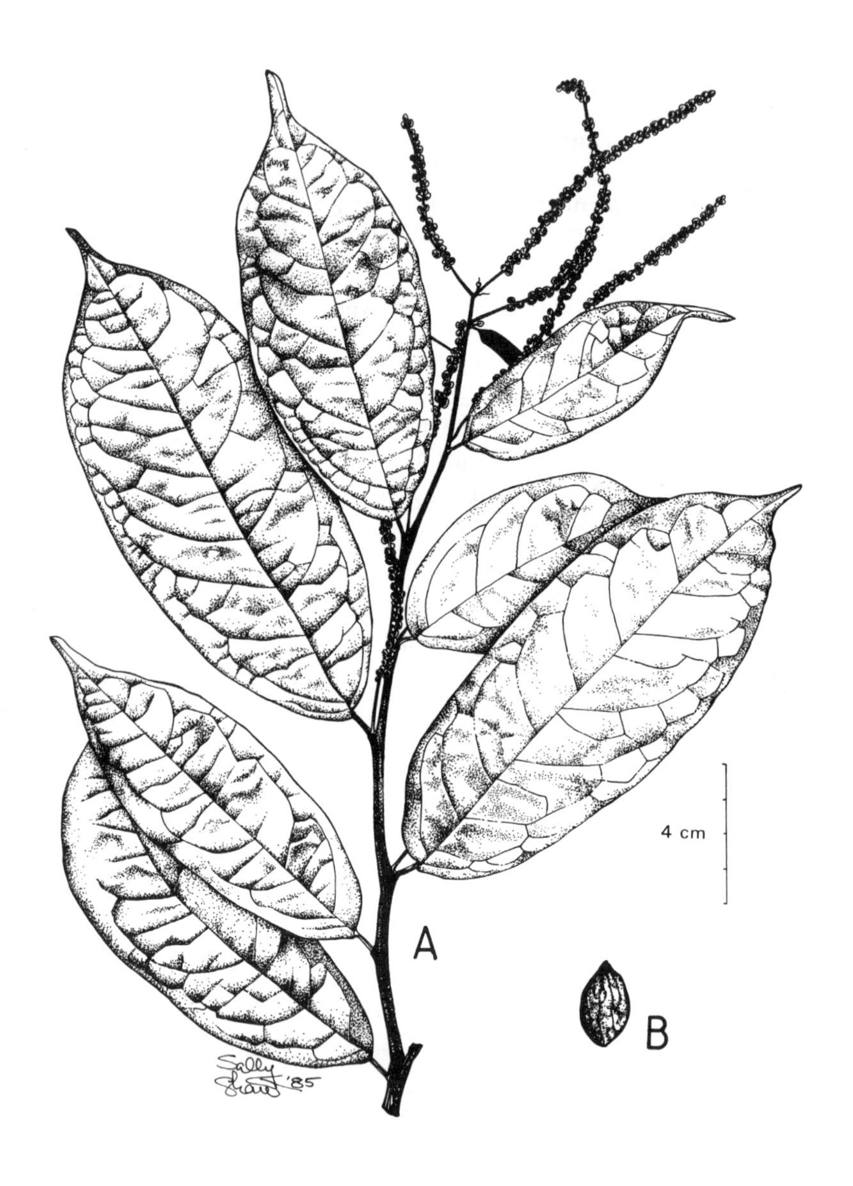

Fig. 31 *Rhyticaryum longifolium* K. Sch. and Laut.
A Habit diagram, **B** Fruit.

The leaves are cooked and eaten in the Solomon Islands.

Chemistry. No alkaloids are present in the leaves or the bark (Hartley, 1973).

Local name	Village name
Nangoka	Aseki

LABIATAE

Ocimum basilicum L. (Fig. 32) WEI-M-01
Basil

Description. A very aromatic, lemon scented herb, erect, branched, from 0.5–1 m high. Corolla white, pinkish or violet. The fruit is dark brown, nutlets 1.5 mm.

Ecology. Found in settled areas and in open waste places, roadsides, teak forests, dry paddies, up to 450 m in New Guinea, once found at 1150 m. Flowers year round.

Distribution. Distributed throughout the tropics and throughout Malesia.

Medicinal use. In Manki, the leaves are heated over an open fire and then pressed against the area of an enlarged spleen. In Nasingalatu, the whole plant is heated over a fire and the juice is squeezed onto sores.

In other parts of the country, the uses vary: in Central Province, the plant is boiled and a person with fever inhales the steam. Also, the leaves are chewed to relieve toothache. On the Papuan coast, the leaves and flowers are used to combat gonorrhoea. The leaves are used in a decoction as a carminative and a stimulent and as a remedy for coughs, in washing ulcers and for hiccups. The flowers are used to treat the coughs of young children. The nutlets are said to have stimulant and diuretic properties. They are also used as an aphrodisiac and for gonorrhoea, dysentery, constipation and sore eyes. The roots are used for bowl complaints and a febrifuge. Basil is used widely as a condiment.

Chemistry. The essential (volatile) oil has been extensively studied. The components include 1,8-cineole, methylchavicol,

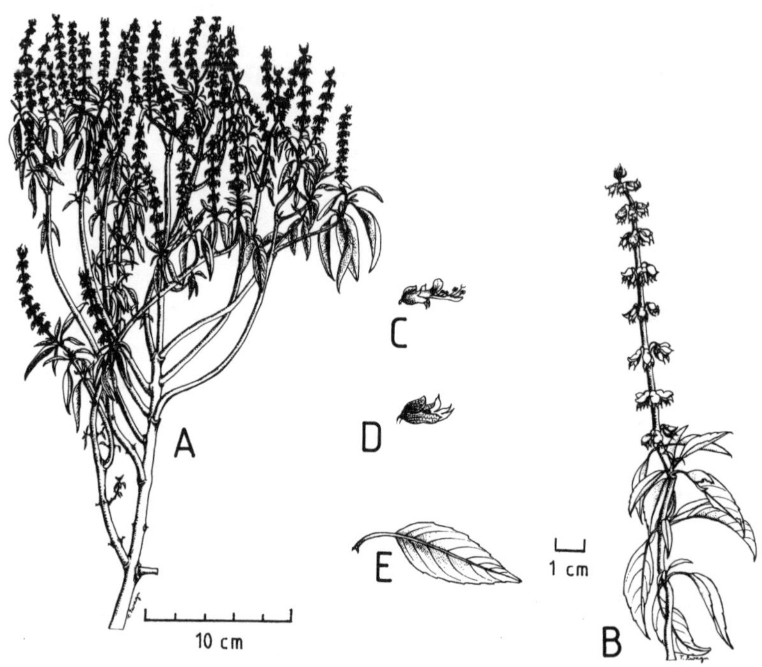

Fig. 32 *Ocimum basilicum* L.
A Habit diagram, **B** Inflorescence arrangement,
C Flower, **D** Seed arrangement, **E** Leaf.

linalool, eugenol, ocimene, safrole, anethole (Duke, 1985). Cineole is effective for treatment of bronchitis, laryngitis and pharyngitis. It is also an expectorant. Linalool is an anticonvulsant. Eugenol is an analgesic. Eugenol and safrole have anaesthetic and antiseptic properties. Anethole is a carminative and an expectorant (Duke, 1985).

Local names	Village names
Saiweso	Manki
Kembaco	Nasingalatu

Plectranthus congestus R. Br. WEI-N-05
Coleus

Description. Herbaceous, up to 1.2 m tall. Leaves round-ovate, crenate margins. Inflorescence branched, up to 70 cm long, flowers white or blue, aromatic.

Ecology. Widely cultivated in gardens.

Distribution. No information.

Medicinal use. In Nauti, the whole plant is soaked in fresh water and used to bathe a person suffering from malaria or fever.

In the highlands, the leaves are squeezed and rubbed on scabies. In the Sepik, the juice from squeezed leaves is rubbed on small sores.

Chemistry. No information could be found.

Local name	Village name
Kawa	Nauti

Plectranthus scutellarioides (L.) R. Br. WEI-N-01

Description. Herb, ca. 1 m tall. Leaves ovate, acute, often purple beneath. Flowers in false spikes, 10–30 cm long. Corolla white or blue, tubular.

Ecology. Common. Found wild in bush margins and grassland. Cultivated in gardens.

Distribution. The leaves of the cultivated plants are variable in shape, margin, colour and amount of pubescence.

Medicinal use. In Nauti, the stem is chewed and spit onto wounds inflicted by spear, knife or axe. The spittle is also exhaled through the nose to treat sore nose, nose bleeding and headache.

In Simbu Province, hot young leaves are squeezed and rubbed on cuts and sores. They are also used as baby napkins. In North Solomons Province, crushed leaves are rubbed on pimples, grille and leprosy.

Chemistry. No information could be found.

Local name	Village name
Miango	Nauti

<u>*Pogostemon cablin* (Blanco) Benth.</u> WEI-N-07

Description. Herb. Erect, aromatic. Stem and branches tomentose. Leaves are ovate, with toothed margins. The flowers are arranged in a raceme forming a terminal panicle. The fruits are small, black shiney nutlets.

Ecology. Garden, lands and clearings and in settled areas up to 1900 m.

Distribution. Sri Lanka and continental SE Asia.

Medicinal use. In Nauti, the leaves are rubbed onto the abdominal area to relieve stomache pain.

Chemistry. Steam distillation of dried, non-fermented leaves yields patchouli oil which is used commercially in perfumes. The main component of the oil is the sesquiterpene alcohol, patchoulol. Present in smaller concentrations is norpatchoulenol (Baver, 1985). *P. cablin* is used in traditional Chinese medicine to treat fungal infections. The alphapyrone derivative pogostone was shown to be the active anti-fungal agent (Farnsworth, 1980).

Local name	Village name
Amtianga	Nauti

LAURACEAE

<u>*Cinnamomum podagricum* Kosterm.</u> WEI-M-01

Description. Small tree.

Ecology. Found in regrowth, montane and in *Castanopsis* forests.

Distribution. No informatiom.

Medicinal use. In Manki, the bark is chewed and spit on the body of a person suffering from chest pain or a cold. In Bukua, the bark of *Cinnamomum* sp. is scraped, mixed with lime and rubbed on a person possessed by an evil spirit.

A related species, *C. archboldiana,* is used in the Southern Highlands to protect children from sickness by spitting the chewed bark on the child's body.

The Wau Ecology Institute Entrance Offices

Mountainous collection area near the village of Aseki in the Morobe Province of Papua New Guinea

Highlands people near the village of Kaintiba in the Chimbu Province

Coastal area near Buso village in the Morobe Province

Coastal vegetation along the Pacific coast in Morobe Province

"Save" man and woman at Buso village in the Morobe Province

“Save” men at Kangurua village in the Finschhafen area of the Morobe Province

Collecting medicinal plants near Biawen in the Morobe Province

Aid Post hut and orderly at Qua Qua

Medicinal plant found on Mt. Kaindi near Wau in the Morobe Province: *Rhododendron macgregoriae* L. used against tropical ulcers

Inflorescence of the shrub *Scaevola sericea* Vahl. The leaves are used to treat a cough and the sap is applied directly to a sore.

Inflorescence of the medicinal plant *Cassia alata* L. The leaves are used to treat grille *(Tinea imbricata)*.

Highlands man with a “stick” of traditional salt produced from a fern

Chemistry. Alkaloids have been isolated from some species of *Cinnamomum* sp. (Gellert, 1970). Species of *Cinnamomun* in P.N.G. contain large amounts of eugenol and varying amounts of safrole.

Local name	Village name
Kadza ura	Manki

LEGUMINOSAE

Cassia alata L. (Fig. 33) WEI-MB-03
(Gr. name Kassia; alata – winged) WEI-BI-184

Description. A shrub or herb, 2–3 m tall. Leaves pinnate with 9–13 pairs of oblong to obovate leaflets. Flowers in terminal racemes, corolla golden yellow. Fruit a 15–18 cm long pod, with a leaflike wing on each valve.

Ecology. Common and widespread. Found in rain forest and sago swamp edges and regrowth and roadsides. Sea level to 1000 m.

Distribution. Of South American origin but now pantropical.

Medicinal use. The leaves are widely used throughout the country as a traditional remedy to treat grille (*Tinea imbricata*). In Mundala, the leaves are crushed and rubbed on the infected skin.

It is also used to treat ringworm in Malaysia and Indonesia (Burkill 1966).

Chemistry. The bark contains small amounts of alkaloids (Hartley, 1973). An anthraquinone is also present (Tutin, 1911). The fruit and leaves of *C. senna* contain anthracene derivatives of which sennosides A and B are the principle active constituents.

Local names	Village names
Agla	Mundala
Gala	Keregia

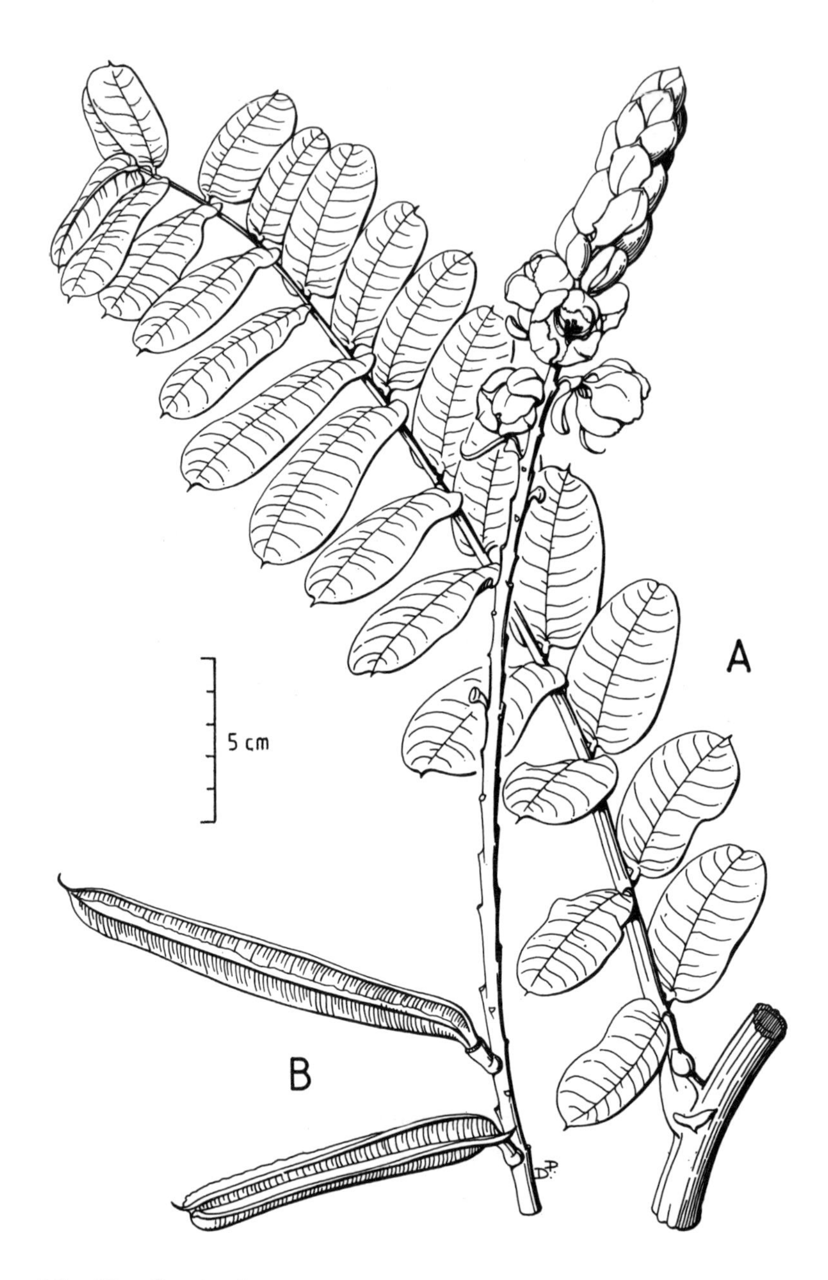

Fig. 33 *Cassia alata* L.
A Leaf arrangement, **B** Fruit and inflorescence

Caesalpina L. sp.

WEI-K-08

Description. Trees, shrubs or woody climbers. Leaves are large and bi-pinnate. Flowers yellow in axillary racemes. Ten stamens.

Ecology. No information.

Distribution. About 40 species, all tropical.

Medicinal use. The stem sap is collected and drunk to treat pneumonia.

Chemistry. No information could be found.

Local name	Village name
Muzoro	Kangarua

Erythrina variegata L.

E. indica

(Gr. erythros – red)

Description. Tree, from 10–30 m tall. The trunk and branches densely prickly. Leaves pinnate, petioled (12–30 cm long). Flowers in racemes, in whorls of 9–12. Fruit a beaked, cylindrical pod, 12–18 cm long and constricted between the seeds.

Ecology. A common coastal species and sometimes planted as a living fence. Found on the coast, monsoon areas, dunes, scrub and regrowth; from sea level and cultivated up to 900 m. Leaves are edible.

Distribution. Origin unknown. Occurs in Tanzania, Indian Ocean Islands, China, Taiwan, India, Malesia and parts of Polynesia.

Medicinal use. In the Finschhafen area, it is used in the treatment of boils; the leaves are crushed and placed on the boil to hasten the removal of pus. Also, the root sap is drunk to treat side pains and the outer skin of the stem is wrapped in a leaf, heated over a fire and placed on a sore daily until healed.

Parts of this tree have various uses throughout the country. In Milne Bay, the bark is used for sores and swellings, in Central Province, the roots are used to treat bronchitis, in the Sepik, the leaves are used to treat coughs. In North Solomons Province, the leaves are used to ease stomach ache and in New Britain, the leaves are rubbed on the head of a person with fever.

This species has a wide range of uses in many other countries, ranging from liver ailments in China (Perry 1980) to a febrifuge in Australia (Webb 1948).

Chemistry. Hydrocyanic acid had been detected in the leaves and bark; these contain saponin and erythrinine, which acts as a nervous depressant. The alkaloids erysovine and erysodienone are also present in the bark (Hartley, 1973). The seeds are poisonous, containing saponin and an inert alkaloid (Morton 1971). Five isoflavones have been isolated from the bark of this species.

Local names	Village names
Ben	Quaqua
Beng	Keregia
Ba	Nasingalatu
Namatia	Buang

Flemingia strobilifera (L.) R. Br. ex Ait. (Fig. 34)

Description. Erect shrub, from 0.6–2 m high. Twigs with white hairs. Leaves entire, ovate to oblong-ovate, petioled. Flowers in panicles or racemes, each flower enclosed by a greenish-yellow bract. Fruit an ellipsoid pod, 2-seeded.

Ecology. Found in grassland, savanna, scrub and plantations; also in regrowth from 5–1300 m.

Distribution. Occurs from Asia to China and Malesia; widely grown in the tropics and extensively naturalized in the Pacific.

Medicinal use. In Keregia, the flowers and leaves are rubbed on the legs of babies who are slow in learning to walk. Children who are slow in talking are given a similar treatment in the mouth.

In New Britain, the seeds may be chewed as a contraceptive and the leaves are used to wash a baby after it is born.

Used as decorations for sing sings and as an ornamental shrub.

Chemistry. The glycosides, phloridzin and naringin have been isolated from the leaves (Saxena, 1976). Naringin has anti-inflammatory properties (Duke, 1985). Many chalcones, flavanones, iso-

Fig. 34 *Flemingia strobilifera* L. (R. Br. ex Ait)

flavones and flavanoids have been obtained from *Flemingia* species (Krishnamurty, 1980).

Local name	Village name
Aafec	Keregia

Pterocarpus indicus Willd. (Fig. 35) WEI-Sq-06
(Gr. pteron – wing; karpos – fruit)

Description. A large tree, to 30 m tall, with distinctive round leaflets and fragrant flowers. Flowers gregariously, decidious, has a rapid growth rate.

Ecology. Shoreline, forest, rain and secondary forest. Sometimes riverine, also in plantations. Wild by the sea and along tidal creeks and rivers by the coast, from sea level to 750(–900) m.

Distribution. Occurs in Southeast Asia, Indo China, Pacific Islands, throughout Malesia.

Medicinal use. A decoction of the leaves is drunk to treat dysentery in Suquang and the red resin from the bark is mixed with water and drunk to treat amaemia in Keregia.

The bark is used to treat pneumonia on Manus Island, the leaves are chewed to ease a bad cough on Karkar Island. In New Britain, the leaves are used to treat stomach ache and in Northern Province, the bark sap is drunk to stop diarrhoea. In Central Province, the leaves are used to treat malaria and wounds and the flowers are used to relieve headaches. In Milne Bay, the juice from the bark is used on sores, and in North Solomons Province, the leaves are used to treat headaches. In addition to the great medicinal value of this species, it is unfortunately one of the most important timber trees in Papua New Guinea.

Used as a living fence and is also traditionally worn strung around the waist.

Chemistry. Alkaloids are not present in the leaves or the bark (Hartley, 1973). However, sapogenins are present in the leaves (Blunden, 1981).

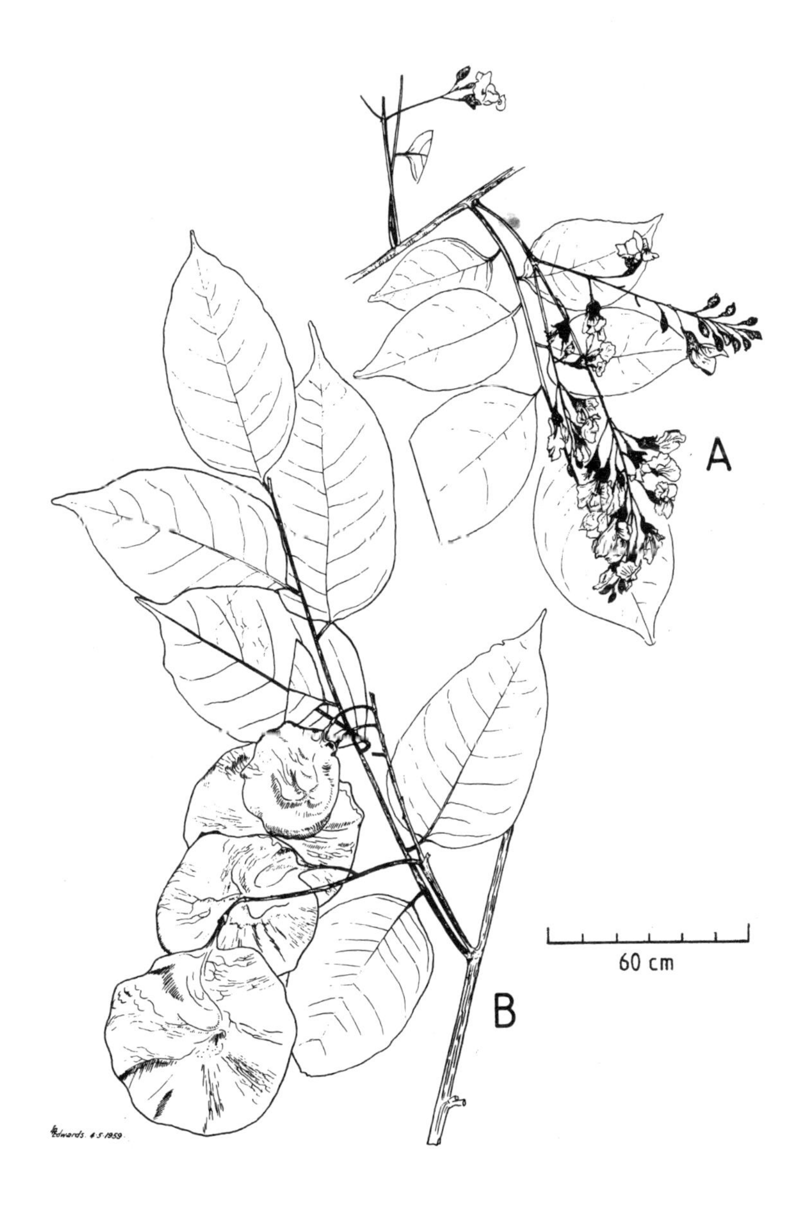

Fig. 35 *Pterocarpus indicus* Willd.
A Inflorescence arrangement, **B** Fruit arrangement.

Local names	Village names
Kinagi	Nauti
Amaurakara	Manki
Pingho	Mapos
Bahink	Wapo
Kamac	Mabsiga
Kamac	Nasingalatu

LINDSAYACEAE

Lindsaya repens (Bory) Thw.

Description. Fern; fronds once-pinnate. Pinnae triangular, crenate on the upper margins. Sporangia on the margins of the teeth.

Ecology. Found creeping on tree trunks; especially on Sago palms.

Distribution. No information.

Medicinal use. In Aseki, the leaves are chewed with traditional salt to treat dysentery. They may be taken regularly until cured. If a couple desire a baby boy, the pregnant woman chews the leaves regularly with traditional salt.

Chemistry. Some species of *Lindsaya* have yielded a diterpenoid glycoside, lindsaeic acid, creticoside A, trans-cinnamic acid and p-hydroxy-trans-cinnamic acid (Satake, 1980).

Local names	Village names
Uyaiwa	
Takwomnga	Aseki

LOGANIACEAE

Fagraea bodenii Wenh.

Description. Tree, up to 30 m, or an erect shrub. Sometimes epiphytic. Leaves opposite, petioled. Inflorescence cymose, 2–15 flowered. Corolla tubular, deep olive outside and cream to white inside. Fruit orange, with a beak.

Ecology. Found in forests and shrubberies, on slopes on limestone hills, from 80–2840 m. Flowering occurs from October to February and May to August.

Distribution. Occurs only in New Guinea.

Medicinal use. To cure an enlarged spleen caused by malaria, the leaves are chewed with traditional salt. Worriers also chew the leaves as a stimulent before battle.

Chemistry. Alkaloids are not present in the leaves or the bark (Hartley, 1973).

Local name	Village name
Pehea	Ascki

MALVACEAE

Abelmoschus manihot L. (Fig. 36)
Hibiscus manihot L.
(Arab. abuelmosk – father of musk)
Aibika

Description. A perennial herb or undershrub, 1–5 m tall. Leaves 5–7 lobed. Corolla large, white or sulfur yellow with a small dark purple centre.

Ecology. ssp. *manihot* – cultivated as a vegetable mainly in SE Asia, also in temperate Europe, as an ornamental, sometimes escaping from cultivation. In Malesia, cultivated particularly in the Sulawesi and the Moluccas. It is a cultigen.

Var. *tetraphyllus* – found from sea level up to 400 m, in areas subject to an annual dry season.

Distribution. In East Java, Lesser Sunda Islands, SW Sulawesi, Philippines, Moluccas, East New Guinea and New Ireland.
forma *leptodactylus* – cultivated. Native of China, now cultivated in all tropical countries.

Leaves eaten boiled as a vegetable.

Fig. 36 *Abelmoschus manihot* L.
$\mathbf{A_{1/2}}$ Leaves, **B** Inflorescence arrangement.

Fig. 37 *Hibiscus rosa-sinensis* L.

Medicinal use. In Buang, the leaves are made into a soup with pitpit (*Saccharum edule*) and used to treat asthma. The glutinous sap is used to treat a cough or bad cold. In many areas, pregnant women frequently consume aibika soup to ensure a healthy baby.

In North Solomons Province, the leaves are squeezed and mixed with sugar cane leaves and drunk to promote labour.

In Central Province, the leaves are boiled and the solution is used to bathe a person with a skin rash and in the Southern Highlands, the leaves are boiled and eaten to treat dysentery.

Similarly, in Indochina (Perry 1980), the flowers are used to combat dysentery.

Chemistry. The related species *A. esculentus* contains an indole alkaloid (Manske, 1968).

Local name	Village name
Dahang	Buang

Hibiscus rosa-sinensis L. (Fig. 37) WEI-Sq-02
(Gr. hollyhock)

Description. Shrub, 2–4 m tall. Leaves ovate, coarsely toothed. Large, showy, colourful flowers.

Ecology. Cultivated throughout the world, largely for the flowers, in the tropics and subtropics.

Distribution. The origin of the species is uncertain, but it is probable it is a native of East Africa, possibly Tanzania, where a related species, *H. shizopetalus* was collected from the wild.

Medicinal use. In two villages in the Finschhafen area, the roots and leaves are crushed and the juice is drunk to treat diarrhoea. In Suquang only, additional uses include the treatment of headache and irregular periods. In addition, the juice is applied directly to sores.

In other parts of the country, the plant is used to treat sore eyes, stomach ache (Central Province), and to induce labour (North Solomons Province and Northern Province).

The plant is used ease childbirth in Somoa (Uhe 1974) and in Fiji (Zepernick 1972). In Indonesia, it is used as a purgative, an abortifacient and to regulate menstruation (Burkill 1966).

Chemistry. Extracts of the flowers have anti-fertility activity, however the active constituents have not yet been identified (Singh, 1982).

Local names	Village names
Kangalu	Suquang
Hangarou	Keregia

Sida acuta Burm. WEI-Q-21

Description. Erect sub-shrub, 1–1.5 m tall. Leaves lanceolate to linear, toothed. Yellow flowers axile, single or in pairs.

Ecology. Common weed of plantations, pasture and roadsides.

Distribution. No information.

Medicinal use. In Quaqua, the roots are crushed and squeezed and the juice is drunk to treat dysentery and diarrhoea.

In Central Province, the leaves are used to relieve dysentery.

In the West Indies, the roots are used to aid digestion and to reduce fever (Ayensu 1981) and in Palau, the leaves are used to relieve stomach ache (Perry 1980).

Chemistry. No information could be found.

Local name	Village name
Gugulu	Quaqua

MELASTOMATACEAE

Medinilla crassinervia Bl. WEI-Wau-185

Description. Shrub or climber, to 15 m high. Fruit yellow-green.

Ecology. Found in montane forest.

Distribution. Occurs Malay Peninsula, Borneo, Sulawesi, Moluccas, New Guinea.

Medicinal use. In Zafiruo, the inner bark is squeezed and a half cup of the sap is given daily to an anaemic person for two weeks, along with a diet of plenty of fresh vegetables. In Aseki, the leaves are chewed with traditional salt to counter the contraceptive properties of the plant *Solanum* sp., to allow conception to take place. The plant is also used to treat nose cancer and is chewed with traditional salt for application to ulcers.

Chemistry. Alkaloids are not present in the leaves or the bark (Hartley, 1973).

Local names	Village names
Mucbalong	Zafiruo
Guiyaya	Aseki

Medinilla teysmannii Miq.

Description. Shrub. Fruit snowy white.

Ecology. Disturbed forest.

Distribution. Occurs in N. Sulawesi, Philippines, Talaud, Moluccas, New Guinea.

Medicinal use. Similar to *M. crassinervia*, the leaves of this shrub are heated over a fire until soft and the juice is squeezed onto tropical ulcers.

Chemistry. Alkaloids are not present in the leaves or the bark (Hartley, 1973).

Local name	Village name
Opata	Aseki

MONIMIACEAE

Palmeria hooglandii Phil.

Description. Woody liana, to 22 m high, branches and foliage glabrous. Leaves opposite, petioled. Blade oblong-elliptic, margin entire. Inflorescence axillary and terminal, in panicles, unisexual. Male inflorescence larger than female; stamens ca. 18. Flowers creamy, scented. Fruit globose, beaked. Achenes purple-black on a red torus.

Ecology. Found in primary and secondary rain forest between 1950–2000 m.

Distribution. Occurs in Papua New Guinea.

Medicinal use. In Morobe, the leaves are chewed, mixed with traditional salt and the mixture is dripped into the nose to treat influenza and coughs.

Chemistry. No information could be found for this species. However, the related *Palmeria gracilis* bark contains the alkaloids laurotetanine and N-methyllaurotetanine (Johns, 1967, 1970). The leaves and stems of *P. arfakiana* also contain lauretetanine and N-methyllaurotetanine (Johns, 1970). These two alkaloids and another, laurolitsine, were found in the bark of a different unspecified species of *Palmeria* (Johns, 1970).

Local name	Village name
Kivika	Aseki

MORACEAE

Ficus baeuerlenii King

Description. Climber. Leaves hairy on the veins beneath, buds and young petioles very hairy. Figs 2–3 cm across, red at maturity.

Ecology. One collection recorded from a stream bank in an open area, at 450 m.

Distribution. No information.

Medicinal use. In a Finschhafen area village, the inner bark is scraped, squeezed and drunk to treat diarrhoea and dysentery. The dose may be repeated again in 2–3 days time.

In other provinces in PNG, related species are used for the same purpose: in New Britain, *F. pachystemon* and *F. wassa* are used to treat diarrhoea and in Northern Province, *F. septica* is used in the treatment of diarrhoea and dysentery. In Malaysia, the related species, *F. variegata*, is also chewed to cure dysentery (Burkill 1966).

Chemistry. Alkaloids are not present (Hartley, 1973).

Local name	Village name
ohota	Zafiruo

Ficus benjamina L. (Fig. 38)

(Gr. sykon – fig tree)

Description. Medium-sized to lofty strangling fig to 9 m, with a fine-leafed crown and drooping branches. Leaves elliptic, leathery, acuminate. Figs sessile on leaf twigs, mostly paired in the axils, ripening orange-red to purple.

Ecology. Scarce in the forest, common as strangling epiphytes on roadside trees.

Distribution. Occurs throughout Malesia to the Solomon Islands and North Australia, India and South China.

Medicinal use. In Buang, the root scrapings are rubbed on the knees of a weak person as magical words are spoken.

In Indochina, the latex is mixed with alcohol and is used for treating shock (Perry 1980).

Chemistry. Alkaloids are not present in the leaves or the bark (Hartley, 1973; Holdsworth, 1983). Abscissic acid is present in the leaves.

Local name	Village name
Nong	Buang

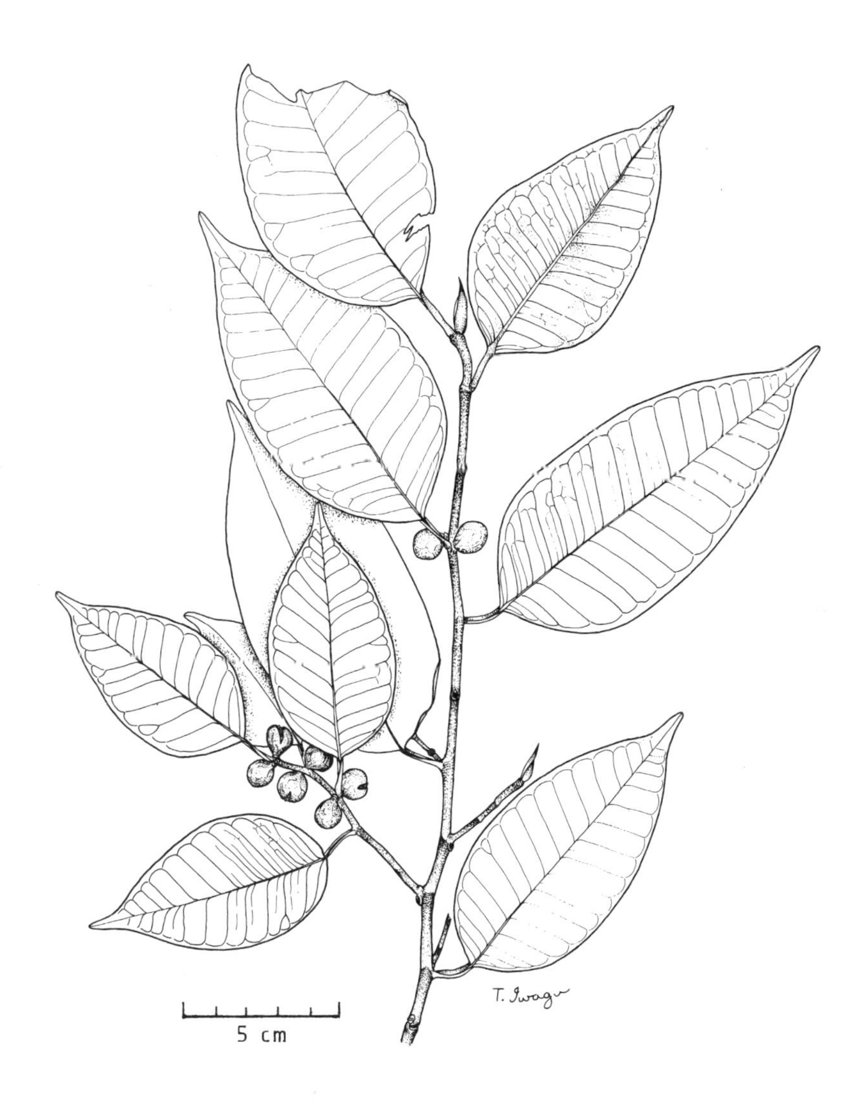

Fig. 38 *Ficus benjamina* L.

Ficus bernaysii King

Description. Small tree, to 6 m. Figs small, to 1 cm across on slender stalks from short, sometimes branching, spurs. Leaves rhombic, elliptic. Sap milky.

Ecology. One collection recorded (in PNG) from 1350 m in secondary forest.

Distribution. No information.

Medicinal use. In a Finschhafen area village, body pains are eased by a massage with leaves that have been heated in a banana leaf over a fire. This treatment is repeated about three times.

Chemistry. Alkaloids are not present in the leaves or the bark (Hartley, 1973).

Local name	Village name
Diwa	Bolinbaneng

Ficus nasuta Summerh.

Description. Liana. Leaves usually markedly cordate, acute. Fruit (fig) to 2.5 cm across, orifice strongly prominant, patent bracts to 4 mm long at base.

Ecology. Found in regrowth.

Distribution. No information.

Medicinal use. In Aseki, the leaves are eaten with traditional salt by women who want to increase their fertility. In one Finschhafen area village, a half cup of bark sap is drunk to treat asthma and other respiratory problems.

Chemistry. Alkaloids are not present in the leaves or the bark (Holdsworth, 1983).

Local names	Village names
Imda	Aseki
Hohoya	Zafiruo

Ficus septica Burm. f. (Fig. 39) WEI-MB-01, WEI-Sq-03

Description. Shrub or small tree, 3–6 m tall. Fruit greenish-white, ribbed, up to 2.5 cm in diameter. Figs flattened-globose, with 10 distinct longitudinal ridges. No latex. Leaves drying green.

Ecology. Found in low secondary growth, rainforest, savannah, old garden clearings, from sea level to 1500 m.

Distribution. No information.

Medicinal use. In the Finschhafen area, the fruit sap is applied to sores and the leaves are used for bathing (in Mundala and Suquang) and in Keregia, the boiled leaves are cooled and the solution is rubbed over the body of a person with malaria or fever.

This tree has various uses in an additional thirteen villages investigated in seven different provinces in PNG (Holdsworth, 1977). The uses include the treatment of coughs in the provinces of Madang, Milne Bay Central and New Ireland, stomache ache in North Solomons Province, diarrhoea, dysentery and body pains in Northern Province, diarrhoea in New Ireland and headaches and fevers in Milne Bay (Normanby Island). On two Islands of Milne Bay Province (Kiriwina and Normanby), the roots are used as a poison antidote.

It also has widespread use in other countries, including the treatment of diarrhoea, stomach ache, wounds and coughs in New Caledonia (Rageau 1973), as a purgative and emetic in the Moluccas (Burkill 1966) and for rheumatic pains with fever in Indonesia (Perry 1980).

Chemistry. The phenanthroindolizidine alkaloids tylophorine and tylocrebine are present (Herbert, 1972). These two alkaloids are anti-tumour compounds. Tylophorine also has anti-allergenic, anti-asthmatic and anti-rhinitic properties (Duke, 1985).

Local names	Village names
Anda	Mundala
Zizigli	Suquang
Ziqililic	Keregia

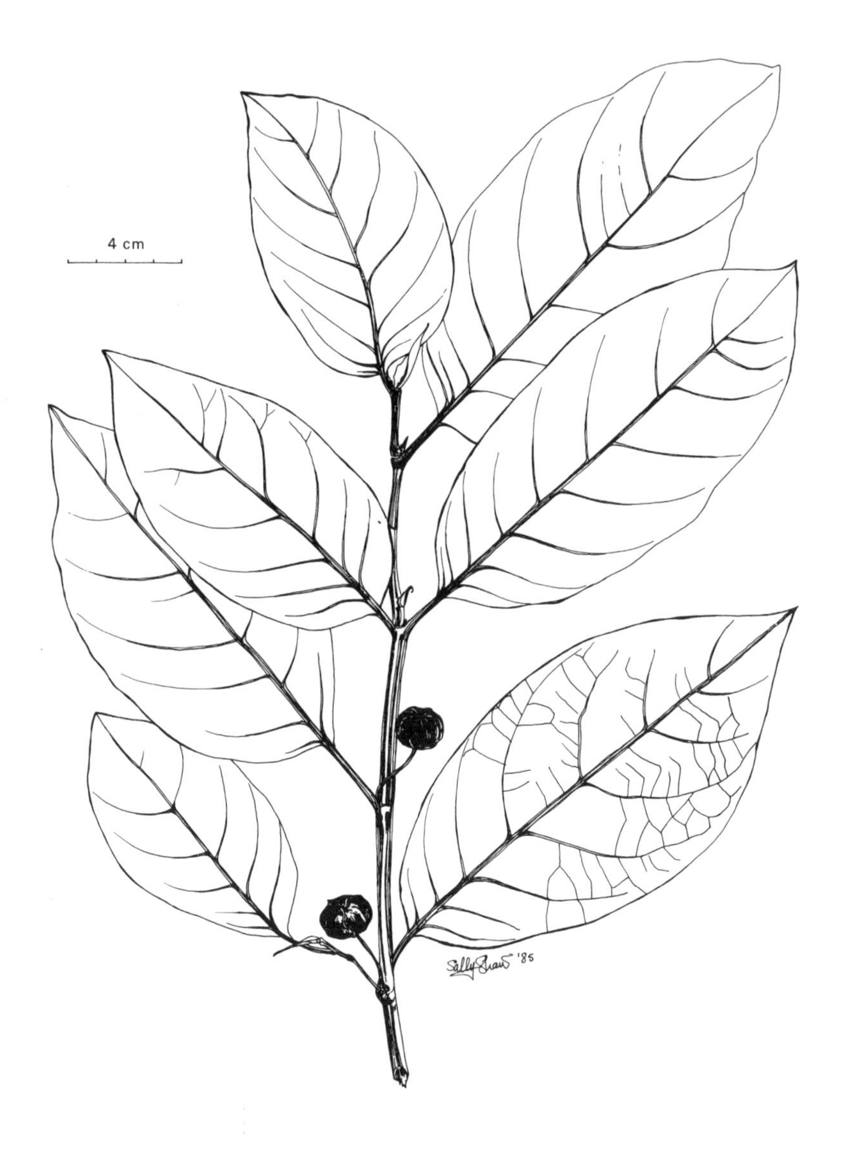

Fig. 39 *Ficus septica* Burm. f.

Ficus subcuneata Corner

Description. Tree to 30 m high. Leaves rhombic or elliptic, base narrowed but obtuse. Figs with peduncles, pubescent with straight hairs.

Ecology. Found in disturbed and lower montane forests.

Medicinal use. In Morobe, the sap of this tree is applied directly to skin rashes and scabies. The treatment is given daily for one week.

Chemistry. Alkaloids are not present in the leaves or the bark (Hartley, 1973).

Local name	Village name
Apiyaka	Ascki

Ficus sublimbata Corner

Description. Figs 2 cm or more across with bracts at base and depressed on stout branches near the base of the trunk.

Ecology. No information.

Distribution. No information.

Medicinal use. The latex is squeezed from the leaves and applied directly to sores in Aseki. This treatment is repeated daily until healing occurs. In Malaysia, several species of *Ficus* are used for the same purpose (Burkill 1966) and in India, the uses include healing sores and leprosy (Chropra 1956).

Chemistry. No information could be found.

Local name	Village name
Feya	Aseki

Maclura amboinensis Bl.

Description. Scrambling, spiny shrub or climber. Leafless; axillary thorns in place of short shoots. Flower heads capitate.

Ecology. Found in lower montane forests.

Distribution. Occurs in Thailand, Malaya, Sumatra, Java, Borneo, Sulawesi, Amboina, New Guinea (var. *amboinensis*). Also in the Philippines and New Guinea as var. *paucinervia.*

Medicinal use. The stem sap is collected and drunk twice daily to treat acute stomach ache and vomiting of blood.

Chemistry. Alkaloids are not present (Holdsworth, 1983). However, other species of *Maclura* have been found to contain xanthones (Deslupande, 1973).

Local name	Village name
Pinambu	Sosoningko

Maclura cochinchinensis (Lour.) Corner WEI-K-09

Description. Climbing shrub, 4–10 m long; thorny. Leaves alternate, acuminate, petiolate. Flowers unisexual, arranged in axillary globose heads. Fruit composite, globose.

Ecology. Found in primary and older secondary forest.

Medicinal use. The stem sap is collected and drunk to treat malaria and pneumonia.

Chemistry. No information could be found for this species, but other species of *Maclura* have been found to contain xanthones (Deslupande, 1973).

Local name	Village name
Qingakum	Kangarua

MUSACEAE

Musa paradisiaca L.

Cooking banana (an interspecific hybrid)

Description. There is a very large mixture of native and mixed cultivars of this species, so that an adequate description is difficult, as all characteristics are quite variable.

Ecology. Plants of tropical humid lowlands; mostly grown between 30 °north and south of the equator.

Distribution. All edible bananas are derived from 2 genera. The centre of diversity of one is in the Malesian area. The other is more abundant in areas with a monsoon climate and a pronounced dry season.

Medicinal use. In a Finschhafen area village, the sap from the sucker plants is pressed onto fresh cuts. Further inland, near Wau, the stem sap is drunk to treat cough, colds and influenza.

In Milne Bay, the soft part of the stem is rubbed on millipede bites.

The related *M. sapientum* is recorded as being used to treat snake bite and other wounds (Futsher 1959); this species is also used in the West Indies to treat worms, snake bites and scorpion stings (Ayensu 1981).

Chemistry. An indole alkaloid is present (Manske, 1968).

Local names	Village names
Umm	Nasingalatu
Lewizikali	Biawen

MYRSINACEAE

Embelia palauensis Mez.

Description. Vine.

Ecology. Found in regrowth.

Distribution. No information.

Medicinal use. In the Watut area, the young leaves are heated over a fire and the juice is drunk to treat an enlarged spleen. In Sosoningko, the freshly squeezed stem sap of *Embelia* sp. is drunk to treat malaria and fever.

Chemistry. The related species *E. ribes* contains the quinone embelin, which has anti-fertility properties (Krishnaswamy, 1980) and anthelmintic properties (Gupta, 1976).

Local names	Village names
no record	Manki
Galohoc	Sosoningko

MYRTACEAE

Octamyrtus pleiopetala (F. Muell.) Diels

Description. Small tree. Leaves oblong-elliptic to lanceolate, petioled. Pedicles and calyx with dense hairs. Numerous stamens. Fruit 1.3 cm diameter, subglobose.

Ecology. Found in montane forest and disturbed areas.

Distribution. Occurs in Moluccas, Irian Jaya, Papua New Guinea.

Medicinal use. The bark is chewed, spit into a bamboo container and traditional salt is added. The mixture is then dripped into the nose to cure colds, fever, sore throat and upset stomach.

Chemistry. Alkaloids are not present in the leaves or the bark (Hartley, 1973).

Local name	Village name
Womnawa	Aseki

Psidium guajava L. (Fig. 40) WEI-Q-14

(psidion, ancient name for unknown plant)
Guava

Description. Shrub or small tree, 3–6 m tall. Twigs 4-angled. Leaves opposite, oblong-elliptic, strongly prominent lateral nerves. Flowers white, axillary. Fruit a berry, yellow when ripe, 6–9 cm long.

Ecology. Native to Brazil. Now naturalized throughout the tropics.

Distribution. Valued for the fruit which has a high vitamin content.

Medicinal use. In two villages in Morobe Province, this species has the same medicinal use: in Quaqua, the juice from the scraped bark is drunk to treat colds and in Aseki, the leaves are chewed with traditional salt to treat colds and influenza. In a Finschhafen area village, the leaves are boiled, sea water is added and the mixture drunk to treat diarrhoea.

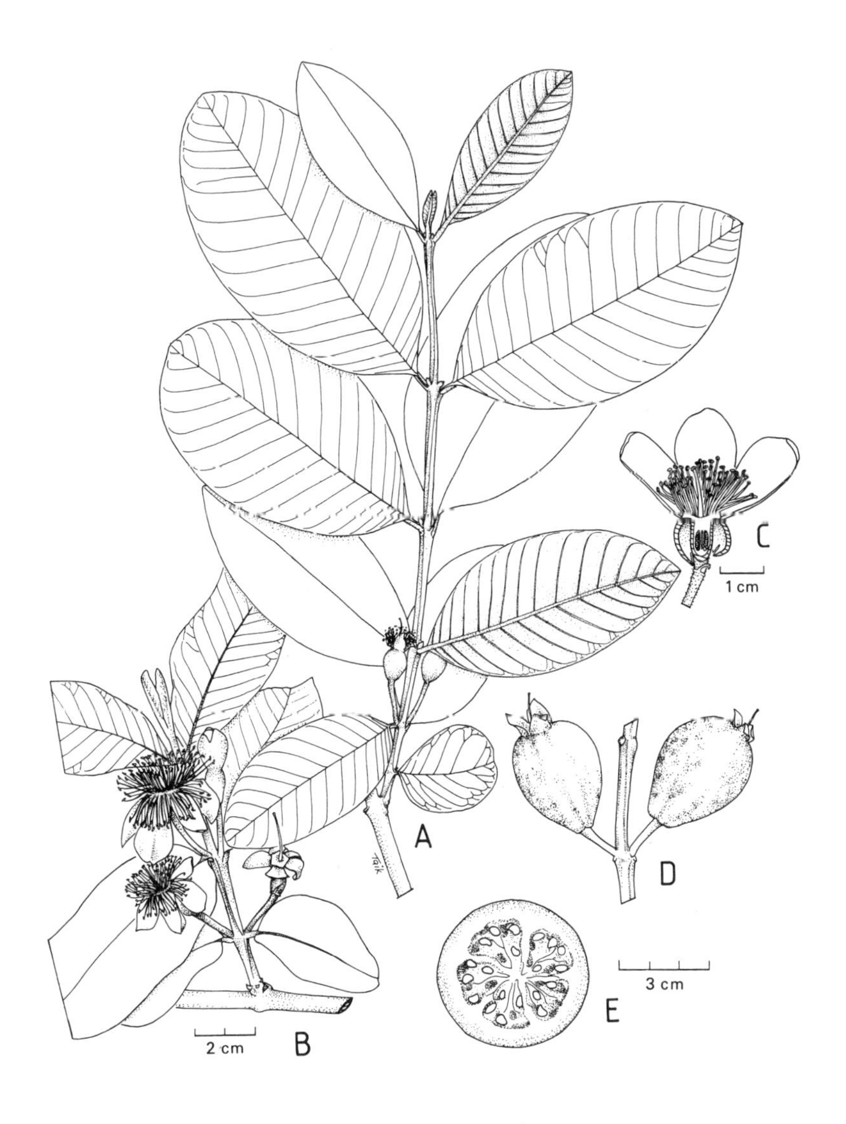

Fig. 40 *Psidium guajava* L.
A Habit diagram, **B** Inflorescence, **C** Flower,
D Fruit habit, **E** Fruit cross-section.

In Central Province, Milne Bay, Karkar Island (Madang Province) and New Ireland, the leaves are used to treat diarrhoea. In Central Province, the leaves are used against malaria and headaches; in New Britain, the leaves are used to treat hepatitis, fever and stomach ache. On Manus Island, the leaves are used to wash scabies and other skin ailments.

This species is used in a number of other countries, in mostly the same way. In India, it is used to treat diarrhoea (Watt et al 1962). In Malaysia, the leaves are used to treat diarrhoea and stomach ache (Burkill 1966) and in the West Indies, the fruit and roots are used against diarrhoea and as a wash for skin diseases (Ayensu 1981).

Chemistry. No information could be found.

Local names	Village names
Guafeng	Quaqua
Guape	Nasingalatu
Watuwa	Aseki

Syzygium malaccense (L.) Merr. & Perry (Fig. 41)

WEI-F-19, WEI-Z-02

Description. Large tree, 10–20 m tall, with a short, flanged trunk, brown flaky bark. Leaves opposite, elliptic. Inflorescences in thick panicles, growing from the wood; flowers showy, pink to dark red. Fruit pink up to 0.6 cm long.

Ecology. Common at lower elevations, in primary forest, regrowth. Widely cultivated in villages for its edible fruit.

Distribution. Native to some part of the Indo-Malaysian region, now more or less pantropical in cultivation.

Medicinal use. In four villages in the Finschhafen area, this tree is used to treat diarrhoea. In the two coastal villages, the sap is squeezed from the leaves and used to treat severe cough. The scraped bark is applied to sores. In the two villages further inland, the bark is crushed, mixed with water and drunk one to two times a day to treat diarrhoea and stomach ache.

In New Ireland, scraped bark is mixed with sea water and drunk to control vomiting and to clear mental stress.

Fig. 41 *Syzygium malaccense* (L.) Merr. & Perry.

Chemistry. Alkaloids are not present in the leaves or the bark (Hartley, 1973). Essential oils are present.

Local names	Village names
Nemuya	Fondengko
Nemuya	Zazaquo
Nemuya	Keregia
Aigang	Nasingalatu

NYCTAGINACEAE

Pisonia umbellifera (Forst.) Seem

P. excelsa Bl.
(after W. Piso, a Dutch artist)

Description. Shrub or tree up to 28 m high. Leaves opposite or in whorls. Inflorescence terminal, branched; a many flowered-umbel. The flowers are funnel shaped.

Ecology. Often found in coastal places, exposed to wind, both everwet and monsoon forest, along riverbanks, creeks, on sand, clay and rocks.

Distribution. Occurs in Madagascar, Mauritius, Mascarenes, Comores, Andaman Is., Hainan, Formosa, Riu-Kiu Is., throughout Malesia, North Australia, Bonin Is., Micronesia, Melanesia (including New Zealand).

Medicinal use. In Morobe, the leaves and bark are burned to ash, mixed with water, then strained into a bamboo container, which is placed over a fire. The solution is slowly evaporated and the remaining crystalline substance is the highly prized traditional salt used in medicine. This salt was once important in barter transactions, bride price and compensation payments.

Chemistry. Alkaloids are not present in the leaves or the bark (Hartley, 1973).

Local name	Village name
Namba namba	Aseki

PALMAE

Areca catechu L.

Common Betel Palm

Description. Tall, slender tree, 10–15 m high. Inflorescence paniculate, multi-branched, with both male and female flowers. Fruit ovoid, 4–6 cm long.

Ecology. Found in primary forest, also planted in villages for the kernel.

Distribution. No information.

Medicinal use. On the Finschhafen coast, the inner seed is chopped, heated over a fire and pressed on sores caused by sea urchins.

A common activity in Papua New Guinea (and Asia), is the chewing of the betel nut with lime and with the leaves or catkins of *Piper betle*. In addition to the stimulant effect, there is also an attributed sedative effect and the mixture may be used to soothe a mad person (Milne Bay). The red mixture is applied to tropical ulcers in New Britain and to sores caused by venereal disease in Northern Province. In New Ireland, the scraped bark is mixed with sea water in a leaf of *Indocarpus fagiferus* and drunk to treat asthma.

In Irian Jaya, parts of this tree are used on wounds, swellings and other skin afflictions (Zepernick 1972). The nut has long been linked to oral and oesophageal cancer (Morton 1978)

Chemistry. The kernel contains alkaloids and tannins. Of the nine alkaloids that have been isolated, the most abundant alkaloid is arecoline, which has antehelminthic properties and is highly toxic (Chopra, 1933; Mujumdar, 1979)

Local name	Village name
M'bu	Mabsiga

Cocos nucifera L. WEI-F-15, WEI-YB-08
(Gr. coccos – berry)
Coconut Palm

Description. Tree; dioecious.

Ecology. A lowland species, but has been planted up to ca. 1000 m. Requires over 1000 mm rain per year, high sunlight and high humidity.

Distribution. The origin is a topic of controversy. It now has a pantropical distribution. The centre of origin of the cocoid palms most closely related to the coconut palms and where most of the species are now found is in NW South America. It is probably of Melanesian origin and was domesticated in Melanesia where it is widely distributed. It is nowhere truly wild. It has been described as nature's greatest gift to man (Burkill 1966).

Medicinal use. In four villages in the Finschhafen area, the palm is used for medicinal purposes. In Fondingko, the young frond is heated over a fire and the extract is applied to fresh knife or axe wounds. In other villages, the use is the same but the stem is heated and the sap used. Also, the juice of the green, unripe coconut is boiled and drunk to relieve diarrhoea.

Various parts of this palm are used to treat diarrhoea (Manus Island and North Solomons Province), stomach ache (East New Britain and Manus Island), sores in Central Province, scabies in Northern Province and to prevent tooth decay in Milne Bay. In Somoa, coconut oil is used as a laxative and to relieve stomach ailments (Uhe 1974).

Chemistry. The volatile oil contains a number of simple ketones and an alcohol (Guenther, 1952).

Local names	**Village names**
Haomu	Fondengko
Hamu	Keregia
Pang	Mabsiga
Nip	Nasingalatu

PANDANACEAE

Pandanus tectorius Soland. var. *novo-guineensis*

Description. Trees, 4–5 m, supported by crowded prop roots, 20–30 cm long. Stem branched into an open crown. Trunk covered with short, thick, scattered thorns. Leaves thick, finely pointed. Fruit heads orange-red, ovoid.

Ecology. Found abundantly in poorly drained forests. Also cultivated.

Distribution. Occurs in New Guinea.

Medicinal use. In a coastal village near Finschhafen, the new root is scraped and boiled in rain water until the solution turns yellow. The cooled liquid is then used daily to treat gonorrhoea.

In New Hannover, the bark is scraped, mixed with water in a wild ginger leaf and the solution is drunk to sedate a mentally disturbed person.

Chemistry. No information could be found.

Local name	Village name
Galeng	Nasingalatu

PIPERACEAE

Piper aduncum L. (Fig. 42) WEI-Sq-05

Description. Monoecious shrub or slender tree up to 8 m. Monoecious. Leaves ovate, 16 cm long, petioled. Inflorescence often bisexual, curved, as long as the leaves. Flowers in dense spirals. Male flowers have 2–3 stamens; female flowers sessile.

Ecology. An aggressive weed found on roadsides or in bushland, most often near cultivation.

Distribution. Common throughout a large part of tropical America. Occurs in Mexico, Central America, Northern South America, West Indies. Naturalized in many places in Malesia.

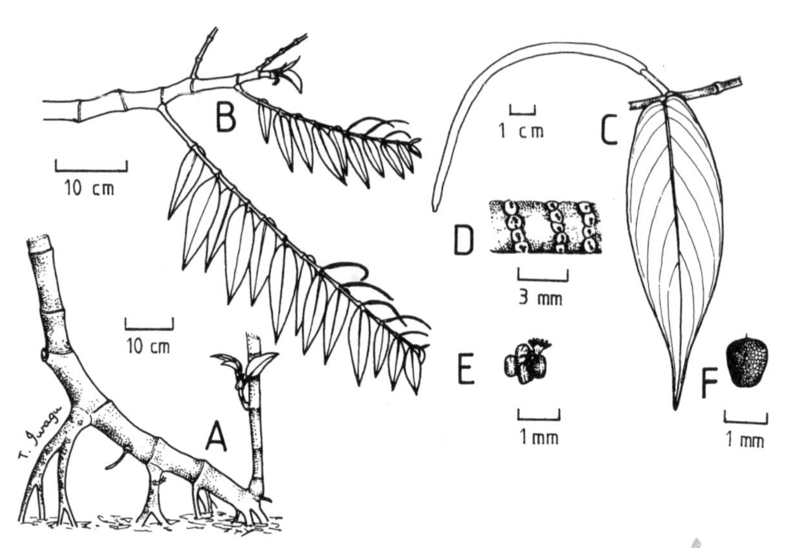

Fig. 42 *Piper aduncum* L.
A Root habit, **B** Branch habit, **C** Leaf and inflorescence, **D** Detail of inflorescence arrangement, **E** Inflorescence, **F** Fruit.

Medicinal use. In three villages in the Finschhafen area, the plant is used on fresh cuts: in Suquang, the stem and bark are scraped and placed on fresh knife, axe or spear wounds, and in the other two villages, the top leaves are crushed and applied to the wound. The new leaves are also used as a bandage. The treatment may be repeated daily as necessary.

Chemistry. Alkaloids are not present in the leaves, bark or the root (Hartley, 1973).

Local names	Village names
Karawe	Suquang
Garoac	Keregia
Kili	Nasingalatu

Piper betle L. (Fig. 43)
Betel pepper

Description. Climber, leaves 5–9 nerved, acuminate. Dioecious. Flowers in pendulous, solitary spikes. The fruit is a cylindrical, fleshy, composite of berries.

Fig. 43 *Piper betle* L.
A Habit diagram, **B** Inflorescence arrangement.

Ecology. Epiphytic; widely cultivated throughout the tropics. Found wild in scrub and secondary forest.

Distribution. Occurs from Malaya to India.

Medicinal use. In the Finschhafen area, the leaves are heated and the juice is applied daily to sores.

Widely used in Asia for chewing along with lime and betel pepper; the mixture acts as a stimulent with a possible intoxicating effect. The mixture is spit onto wounds and abcesses in Irian Jaya (Sterly 1970). In the Philippines, the fresh, crushed leaves are used as an antiseptic on wounds (as in Morobe Province) and as a poultice for boils (Padua et al 1978).

Chemistry. Alkaloids are not present (Hartley, 1973). The essential oil contains mainly eugenol and eugenol methyl ether. Other components are estragol, caryophyllene, cadinene, chavicol, cineole, carvacrol, allyl catechol and chavibetol (Duke, 1985). The essential oil possesses activity against several gram positive and gram negative bacteria (Steenis-Kruseman, 1953; Duke, 1985). Eugenol has analgesic, antiseptic and anaesthetic properties. Methyl eugenol is a sedative . Cineole is an expectorant and assists in the treatment of bronchitis, laryngitis, pharyngitis. Carvacrol is an anthelmintic, antiseptic and tracheal relaxant.

Local name	Village name
Bala	Nasingalatu

Piper canicum Bl.

Description. Climber, leaves ovate, 14 cm long, petioled. Dioecious, female inflorescence up to as long as the leaves, not densely flowered. Fruits small, ovoid-globose.

Ecology. Epiphytic; widespread in disturbed and montane forests.

Distribution. Widely distributed in Malesia through to New Guinea to the Solomon Islands and Australia.

Medicinal use. The heated leaves are rubbed onto pimples on the skin.

Chemistry. No information could be found.

Local name	Village name
Mocuc	Nasingalatu

Piper wichmannii C. DC. (Fig. 44)

Description. Shrub or tree with prop roots, soft wooded. Leaves broad ovate, base asymmetrically cordate. Dioecious; inflorescence up to as long as the leaves, female inflorescences usually shorter. Male flowers with 2 stamens.

Ecology. Aborescent growth habit, widespread.

Distribution. Occurs in New Guinea, Bismarck Archipelago, Solomon Islands. Most common species of *Piper* in New Guinea.

Medicinal use. In Buang, the young leaves are rubbed on scabies during bathing.

Many species of *Piper* are used to bathe swellings and sores (Perry 1980). A related species, *P. nigrum*, is used to treat scabies and sores in India (Chopra 1956).

Chemistry. No information could be found.

Local name	Village name
Kapi	Buang

PITTOSPORACEAE

Pittosporum ferrugineum Ait.
(Gr. pitta – resin; spora – seed)

Description. Small tree or shrub, 3–15 m high, (rarely to 20 m). Leaves elliptic, spirally arranged. Inflorescence terminal. Flowers white to pale yellow. Ovary densely rusty hairy. Stamens 5. Fruit a capsule, ripening to orange. Seeds bright red, immersed in a sticky pulp. The bark has a pungent, unpleasant smell.

Ecology. A common species in the tropical montane zones, near seashores, swamps, around rocks, along rivers and in open savannah. It sometimes occurs as undergrowth in rain forests or in

Fig. 44 *Piper wichmannii* C. DC.

secondary forests from lowlands up to 1800 m (once recorded at 2800 m in New Guinea).

Distribution. SE Asia to Australia and Melanesia, through Malaysia and SW Phillipines. It is the most widely distributed species in Malaysia and can always easily be recognized.

Medicinal use. In Buang, the bark is scraped from the root and pressed into the cavity of a decaying tooth to give temporary relief. In Aseki, the bark is chewed and spit into a bamboo container, mixed with traditional salt and dripped on a person's nose to treat stomach ache and enlarged spleen caused by chronic malaria.

In Central Province, the cure for toothache is the same as in Buang and the Eastern Highlands: the bark is shredded, baked in bamboo and eaten with traditional salt and green vegetables to induce vomiting when poisoning has occurred.

In Malaysia, the bruised leaves and fruit are used as a fish poison and the leaves and roots are used medicinally. In the Philippines, the fruit of a related species, *P. resiniferum* is used to treat abdominal pains (Quisumbling 1951).

Chemistry. Alkaloids are not present. The leaves and fruit contain saponin and a triterpenoid (Perry, 1980).

Local names	Village names
Gotubang	Buang
Ukhewa	Aseki

RANUNCULACEAE

Clematis clemensiae Eichler WEI-S-07, WEI-Mb-07
(Gr. klema – climber)

Description. Vine.

Ecology. Found in regrowth.

Distribution. No information.

Medicinal use. In two villages in the Finschhafen area, the leaves are crushed and sniffed to clear a blocked nose and to relieve headache.

The same use is found in three villages in Northern Province and in a village on Manus Island.

Chemistry. Alkaloids are not present (Hartley, 1973).

Local names	Village names
Zamzamfiroro	Suquang
Wamom	Mundala

Clematis papuasica Merr. & Perry (Fig. 45)

Description. Vine.

Ecology. Found in regrowth and grasslands.

Distribution. No information.

Fig. 45 *Clematis papuasica* Merr. and Perry.

Medicinal use. On the Morobe coast near Finschhafen, the leaves are crushed and rubbed on the fungal skin infection, grille. Further inland, in Buang, the leaves are crushed and sniffed to clear blocked nasal pasages and a cold in the nose.

In Central Province, the same practice occurs as in Buang.

Chemistry. Alkaloids are not present (Hartley, 1973).

Local names	Village names
Omwalu	Nasingalatu
Suwone	Buang

RHAMNACEAE

Alphitonia ferruginea Merr. and Pery.

(Gr. alphiton – barley)

Description. Small tree.

Ecology. Common in secondary growth, disturbed areas.

Medicinal use. This plant is used to treat toothache.

Chemistry. The related species *A. petrieri* contains methyl salicylate, a substance commonly used by external application to relieve muscular pain (Webb, 1959). The wood contains several triterpenoid acids and alphitonin (Lassack, 1983).

Local name	Village name
Piheng	Buang

Gouania Lam. sp. WEI-Sq-13

Description. Shrub or slender woody climber. Leaves alternate. Flowers very small, in cymes on long slender spikes. Five petals. Five stamens. Seeds are very shiny black.

Ecology. No information.

Distribution. No information.

Medicinal use. In two villages of the Finschhafen area, the stem sap is collected and drunk one to two times a day to treat malaria. One cupful of sap is drunk to treat pneumonia, cough and asthma.

Chemistry. Alkaloids are not present (Hartley, 1973).

Local names	Village names
Zizibang	Suquang
Manane	Sosoningko

ROSACEAE

Rubus brassii Merr. et Perry

Description: Straggling or climbing unarmed shrub up to 4 m high. Stems woolly to glabrous. Leaves 5 lobed, petioled, oblong-lanceolate. Inflorescence a lax terminal thyrse, up to 12 cm long. Petals falling early, white. Stamens 45–100. Collective fruit ovoid, orange-red when ripe.

Ecology. Found in forests at altitudes from (180–)600–1525 m.

Distribution. Occurs in the Solomon Islands.

Medicinal use. Elderly people may drink the extracted stem sap as a tonic. One cup may be drunk every other day.

Similarly, in Northern Province, a related species *R. ledermanii* is used as a tonic.

In New Caledonia, the related species, *R. rosaefolius*, is also used as a tonic.

Chemistry. Alkaloids are not present (Holdsworth, 1983).

Local name	Village name
Fapa	Sosoningko

Rubus moluccanus L.

(L. ruber – red)

Description A prickly, straggling shrub, 5–10 m long. Leaves ovate, or palmately 3–5 lobed. Flowers in axillary or terminal racemes. Corolla white; fruit bright red.

Ecology. Found in regrowth and secondary forest.

Distribution. No information.

Medicinal use. In Aseki, the leaves are chewed and the sap swallowed with traditional salt to treat internal pains. In the Finschhafen area, a cupful of the stem sap is drunk daily to combat diarrhoea and dysentery.

In the Eastern Highlands, the leaves are chewed with traditional salt and applied to sores. In New Britain, sap from the young shoots is drunk to stimulate labour. Also in Papua New Guinea, Blackwood (1935) reported the use of the leaves for relief of abdominal pains among the Kukukuku people and Powdermaker (1933) reported the use of leaf extracts as an abortifacient in New Ireland.

In Malaysia (Burkill 1966), the roots are used to treat dysentery and the leaves are used to relieve internal complaints, both similar to the practices in Morobe Province. In Indonesia, the young leaves are chewed to prevent miscarriage (Perry 1980) and the plant is applied to boils (Burkill 1966). In India, the plant is used to induce abortion (Chopra 1956).

Chemistry. Alkaloids are not present (Hartley, 1973; Holdsworth, 1983). The triterpene, rubusic acid, has been isolated (Bhattachrya, 1969).

Local names	Village names
Yakekituwa	Aseki
Fapa	Sililio

RUBIACEAE

Anthocephalus chinensis (Lamk.) A. Rich. ex Walp.

(Gr. anthos – flower; kephale – head) (Fig. 46)

Description. Moderate to large trees, 15 – 30 m tall, with horizontally spreading branches arranged in tiers. Leaves oblong-elliptic to elliptic, glabrous. Flowering heads terminal, solitary. Corolla yellow to orange. Five stamens, inserted in the throat of the corolla. Fruitlets fleshy.

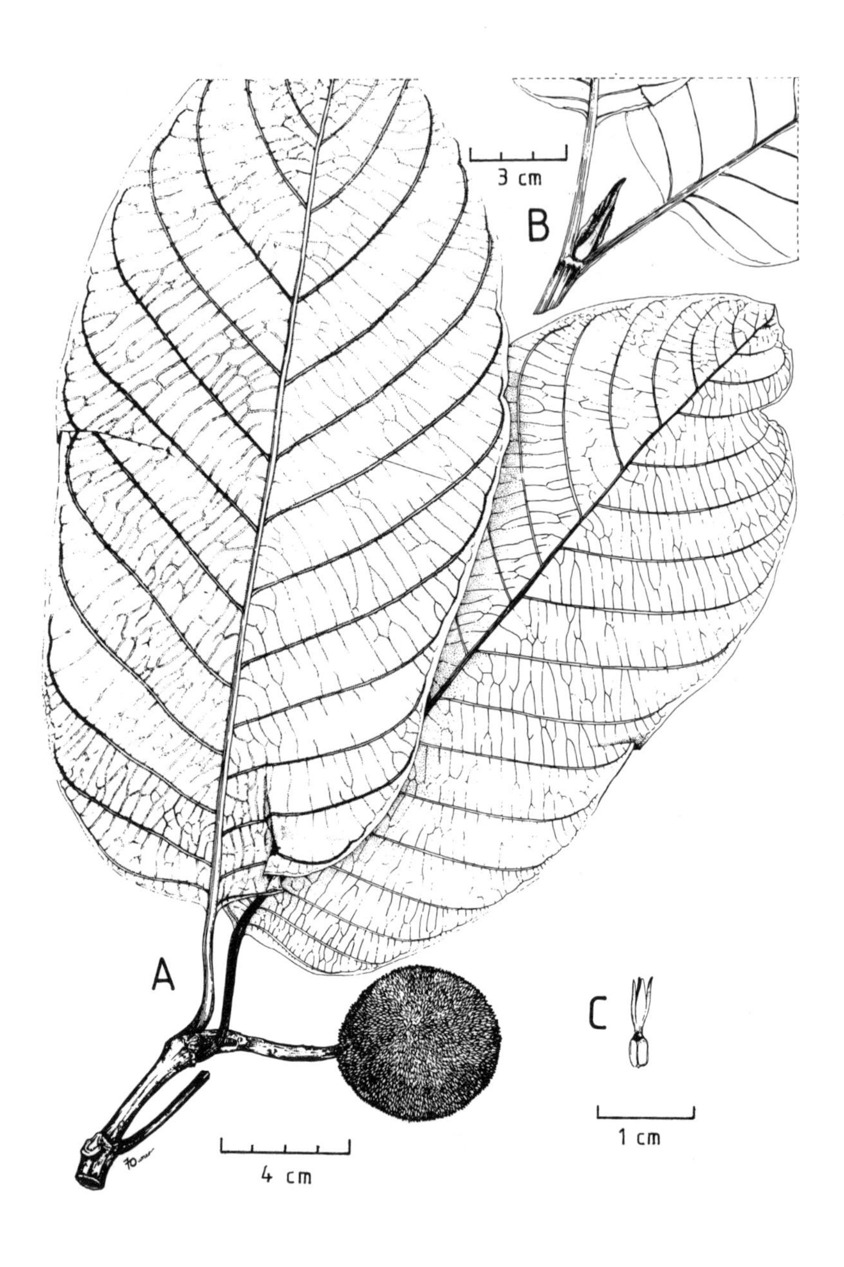

Fig. 46 *Anthocephalus chinensis* (Lamk.) A. Riech. ex Walp.
A Habit diagram, **B** Inflorescence, **C** Flower.

Ecology. In primary and secondary lowland rainforests. Abundant on damp, even swampy ground.

Distribution. No information.

Medicinal use. To treat severe boils, the bark is scraped and mixed with water along with the scraped bark of the tree *Dysoxylum variable*. This solution is drunk daily until the boils heal.

Chemistry. Alkaloids are present in the leaves and the bark (Hartley, 1973; Holdsworth, 1983). The related *A. cadamba* contains 3 indole glycoside alkaloids in the heartwood (Manske, 1979).

Local name	Village name
Gaping	Sililio

Mussaenda ferruginea K. Sch.

Description. Shrub, 3–6 m high, sometimes forming a dense mass. Leaves elliptic. Twigs and flowers rust brown. Flowers in terminal, 3-branched cymes. Corolla tube deep yellow.

Ecology. Found in primary and secondary forest and in regrowth.

Distribution. No information.

Medicinal use. In the mountainous area near Finschhafen, one cup of the stem sap is drunk daily to treat malaria and fever.

In Northern Province, a person with fever is bathed in the cooled extract of leaves boiled in water. Also, headaches are treated by the application of leaves to the head and hair.

Chemistry. Alkaloids are not present (Hartley, 1973; Holdsworth, 1983). The related *M. parviflora* contains iridoid glycosides (Takeda, 1977).

Local names	Village names
Oliticne, Pinambu	Sosoningko

Uncaria ferrea DC.
U. lanosa Wall. var. *ferrea* (Bl.) Ridsd.
(L. uncus – hook)

Description. Climbing shrub, 5–10 m high, stem covered with rust-red hairs. Leaves opposite, long acuminate, oblong elliptic, densely rust-red pubescent underneath. Flowers crowded into globose heads. Fruit a many seeded capsule.

Ecology. Found in regrowth.

Distribution. No information.

Medicinal use. To cure stomach ache, the stem sap is added to vegetable soup, which may be taken weekly until cured. This same treatment is also used for fever, but may be taken more regularly.

Chemistry. The major alkaloids are the oxindoles, pteropidine, speciophylline, isopteropodine and uncarine (Hart, 1967).

Local Name	
Zafengang	Masangko

RUTACEAE

Acronychia elliptica Merr. & Perry

Description. Shrub to small or large tree to 28 m tall. Inflorescence few to many flowered. Flowers yellow to white, drying brown to blackish.

Ecology. Found in primary and secondary rainforests and coastal scrubs from sea level to 2200 m.

Distribution. Occurs in India, Sri Lanka, east to Taiwan and Malesia to Papua New Guinea.

Medicinal use. The leaves are eaten together with traditional salt to treat hookworm after the larvae have been detected in the person's stools.

Chemistry. Some species of *Acronychia* contain acridone alkaloids (Lamberton, 1953).

Local name	Village name
Shimalya	Aseki

SAPINDACEAE

Dodonaea viscosa (L.) Jacq.

(after R. Dodoens, 1517–1585, Dutch botanist)

Description. Shrub or rarely treelet, 1.5–3(5) m tall, glabrous. Leaves obovate, alternate. Flowers greenish-yellow in racemes or panicles. Fruit winged.

Ecology. Found on the coast, mostly on sandy beaches and also in highland grassland.

Distribution. Occurs nearly worldwide, the species is very uniform throughout its distribution.

Medicinal use. Near Finschhafen, the leaves are heated and the squeezed juice is applied to a mother's breasts to increase lactation.

In other highland areas of Papua New Guinea, the heated leaves are used as a poultice for boils, sores and tropical ulcers. Occasionally, the poultice may be used to relieve a sore anus after prolonged dysentery; the person must sit for several hours on the leaves. In the Southern Highlands, the bark is used to treat dysentery. In Simbu Province, the young leaves are applied to sores and cuts.

In India, the leaves are used on wounds, swellings and burns (Chopra 1956). In the Philippines, the bark is used on eczema and ulcers (Quisunbing 1951). In Malaysia, the plant is used to treat flatulence (Burkill 1966). The native people of Australia use the roots to treat cuts and open wounds (Cribb 1981).

Chemistry. Alkaloids are not present (Hartley, 1973; Holdsworth, 1983). The compounds isorhamnetin, hautriwaic acid and a saponin have been extracted (Dawson, 1966).

Local name	Village name
Zalamac	Sosoningko

SOLANACEAE

Solanum moszkowskii Bitter.

Description. A sprawling shrub or woody scrambler, 1–4 m tall. Leaves paired, obovate-elliptic with the base cuneate, apex acumiate. Flowers from the leaf axil. Flowers white and the fruit is a fleshy, bright red berry. The seeds are flattened, oval with a winged margin.

Ecology. Found in montane rainforest, regrowth and disturbed forests.

Distribution. No information.

Medicinal use. The leaves of this vine are chewed with traditional salt as a contraceptive. Also, juice from the crushed leaves is applied to sores.

A related species, *S. verbascifolium*, is used to induce abortion in Malaysia and to treat fevers in Taiwan (Perry 1980).

Chemistry. Alkaloids are not present (Hartley, 1973). However, other species of *Solanum* contain steroid alkaloids which could be starting materials for the commercial manufacture of steroidal drugs (Mankse, 1981).

Local names	Village names
Gotohata, Yakotega	Aseki

STERCULIACEAE

Sterculia L. sp. WEI-B-04
(L. stercus – dung)

Description. Tree or shrub. Black seeds.

Eccology: No information.

Distribution. About 200 species, all tropical.

Medicinal use. In a village in the Wau area, the seeds and roots are crushed and applied to the hair to eradicate head lice. A solution of the crushed plant parts is also used to bathe people afflicted with a cold, asthma or malaria.

On KarKar Island, juice from the young leaves is used to treat malaria and in North Solomons Province, the crushed leaves are applied to ulcers.

Chemistry. Alkaloids are not present (Hartley, 1973). The seeds contain 40–50 % oil, including sterculic acid, a cyclopropene acid (Perry, 1980).

Local name	Village name
Lhelhikomu	Biawen

THEACEAE

Eurya acuminata D.C.
(Gr. eurus – broad; L. acuminata – pointed)

Description. Shrub or tree, to 15 m in height. Fruit globose, glabrous, blue-black. There is a montane form which has thicker leaves and is more hairy.

Ecology. Found in open places from lowlands to the mountains (2000 m).

Distribution. Occurs in Sri Lanka, E. India, SW China, Formosa, Sumatra and Java.

Medicinal use. This plant is used in Aseki, where the leaves are chewed to prevent cough.

Chemistry. Many species of *Eurya* tested for alkaloids have been found to be negative (Hartley, 1973).

Local name	Village name
Yeika	Aseki

ULMACEAE

Celtis hildebrandii Soep.

Description. A large tree up to 45 m, buttressed. Ripe fruits are fleshy, purple or bluish black.

Ecology. In both primary and secondary forests from 0–1000 m, often very common and gregarious locally. Found on various types of soil.

Distribution. Common in the Solomons, In Malesia: Moluccas, New Guinea (west and east, common in New Britain).

Medicinal use. In Aseki, the leaves are chewed with traditional salt and the sap is swallowed to treat internal pain. A person with a sore throat or laryngitis may continue the treatment for two days.

A similar treatment is reported by Blackwood (1935) to be used among the Kukukuku people of Papua New Guinea. In Indonesia, the leaves are chewed to prevent miscarriage (Perry 1980) and Malayans use the plant for internal complaints (Burkill 1966).

Chemistry. No information could be found.

Local name	Village name
Titka	Aseki

UMBELLIFERAE

Oenanthe javanica DC.

Description. Herb or vine.

Ecology. Found in exposed areas.

Distribution. No information.

Medicinal use. In a village in the Watut area, the stem of this plant is chewed to treat cough.

A similar treatment is used in the highlands (Enga Province). In the southern Highlands, the plant is used as an antidote to poison:

with a sorcerer present, the leaves are chewed with traditional salt and wild ginger and vomiting is thus induced. In Central Province, the plant is used to cure headaches.

Chemistry. The related species *O. phellandrium* has medicinal properties; the fruit contains 1–2.5 % essential oil of which about 80 % is the terpene phellandrene (Duke, 1985).

Local name	Village name
no record	Nauti

URTICACEAE

Elatostema Forst. sp. — WEI-B-03

Description. Herb or low shrub

Ecology. No information.

Distribution. No information.

Medicinal use. In the Wau area, the new growing shoots are rubbed on a dislocated knee.

In villages in Northern and Enga Provinces, the leaves are ingested to arrest diarrhoea.

Chemistry. Alkaloids are not present (Hartley, 1973).

Local name	Village name
Silang silang	Biawen

Laportea decumana (Roxb.) Wedd.
Dendrocnide decumana — WEI-N-03, WEI-M-13, WEI-B-12

Description. Perennial shrub, sub-shrub or tall herb, to 2 m high. Leaves rugose, dense with long irritant hairs, young ones dense, woolly; toothed margin. Stem densely armed with long rigid, irritant hairs. Monoecious. Inflorescence unisexual, paniculate, branched, axillary, solitary, male flowers in lower leaf axils, female in upper leaf axils. Male flowers have four stamens, four tepals. Fruit an achene.

Ecology. Widespread; in primary and secondary forests, disturbed areas.

Distribution. Occurs in Borneo, Sulawesi, Moluccas, New Guinea, (also cultivated in India and Java).

Medicinal use. The use of this plant is widespread in Morobe Province and the rest of Papua New Guinea. In Morobe Province alone, it is found in use in 6 villages investigated; generally, the leaves with their stinging hairs are rubbed onto the body to alleviate aches and pains and muscular fatique. A more specific use (in Manki) is rubbing the leaves on the area of an enlarged spleen.

Other uses recorded in thirteen additional villages throughout the country include the treatment of headache, stomach ache, bruises and intestinal pain.

Chemistry. Alkaloids are not present (Hartley, 1973; Holdsworth, 1983).

Local names	Village names
Kauia	Nauti
Ampiorura	Manki
Kalak	Biawen
Zong	Zafiruo
Kauwa	Aseki
Jajap	Buang

Laportea interrupta (L.) Chew.
Urtica interrupta L.

Description. Annual herb, 1–2 m high. Stem woody at base, irritant hairs towards apex. Leaves ovate with irritant hairs, toothed margin, petioled. Monoecious. Inflorescence bisexual, paniculate, up to 30 cm long. Male flowers have 3 – 4 stamens and tepals. Female flowers have 4 tepals. Fruit an achene.

Ecology. A species of herbaceous weeds under partial shade in abandoned gardens, plantations, roadsides. Widespread.

Distribution. Occurs in New Guinea, Congo, E & S Africa, Abyssinia, India, Sri Lanka, Japan, China, throughout SE Asia, Malesia to NE Australia and the Pacific Islands.

Medicinal use. The uses of this species are similar to the related species, *L. decumana*. In Nasingalatu, the leaves are rubbed on an aching head, tired muscles and a sore stomach.

Chemistry. No information could be found.

Local name	Village name
Locgum	Nasingalatu

Nothocnide repanda (Bl.) Bl. WEI-Sq-10

Description. Woody climber or scandent shrub; no irritant hairs. Leaves spirally arranged, petioled, simple, smooth margin, scabrous on the lower side and veined on the upper. Inflorescence spicate, axillary, unisexual flowers. Male flowers have 4 stamens, 4 tepals. Fruit an achene.

Ecology. Found in regrowth, disturbed areas.

Distribution. Occurs in Sumatra, Java, Bali, Borneo, Philippines, Sulawesi, Moluccas, New Guinea, Bismarck Archipelago, Solomon Islands.

Medicinal use. In two mountain villages in Morobe, one cup of the stem sap is drunk daily to ease a sore throat, and the stem sap together with vegetable soup and the warmth of a fire is used to treat influenza or fever. In Suquang, the stem sap from *Nothocnide* sp. is drunk to ease body pain.

On Karkar Island, the sap is squeezed from the new top leaves and applied to sores or rubbed on to the chest to soothe a bad cough. The leaves may also be used as a tonic to speed recovery after sickness. In the Sepik, the leaves are used to cure mouth ulcers.

Chemistry. Alkaloids are not present (Hartley, 1973; Holdsworth, 1983).

Local names	Village names
Garoho	Sosoningko
Galoho	Sililio
Backilo	Suquang

Pipturus argenteus (Forst.) Wedd. WEI-B-70

(Gr. pipto – deciduous; ura – tail; L. argentus – silvery)

Description. Small tree, up to 12 m tall. Leaves alternate, ovate, dark green above, silvery beneath; petioled. Fruit light green, berry-like.

Ecology. Common in secondary forest.

Distribution. Occurs in Australia, Fiji, Tonga, Marquesas, Mariana, Bismarck Islands, New Guinea, Moluccas and Malaysia.

Medicinal use. In three villages as far apart as Finschhafen, Wau and Aseki, this plant is commonly used in the treatment of asthma. The preparation differs slightly, however. In Suquang and Biawan, the inner bark is scraped and mixed with the leaves of *Hibiscus rosa-sinensis* and the solution is drunk to treat cough and asthma. In Aseki, it is the leachate from rainwater falling on the leaves that is used. In Buang, the use is entirely different; the scraped bark is applied to spear wounds to help extract the spear head.

In other areas of the country, the plant has a multiple of uses. In New Britain, the juice from crushed leaves is drunk to relieve fever or headache. In three villages in North Solomons Province, the plant is used in the treatment of cough, stomach ache and centipede bites. On Karkar Island, the roots are squeezed and the juice drunk to treat malaria or severe cough. In the Western Highlands (Stopp 1963), the root juice is applied to wounds or aching teeth. In New Ireland, the sap from the scraped inner bark is drunk to assist a woman in labour.

There are many recordings of the use of this plant in other parts of the world. Uses include the treatment of stomach ache in Somoa (Uhe 1974) and the Solomon Islands (Cribb 1980), where it is also used as an abortifacient (Foye 1976), and boils in the Philippines (Quisumbing 1951).

Chemistry. Small amount of alkaloids are present in the leaves (Hartley, 1973).

Local names	Village names
Lhe	Suquang
Lumbai	Biawen
Yiwiya	Aseki
Lul	Buang

VERBENACEAE

Faradaya parviflora Warb.

Description. Woody climber.

Ecology. No information.

Distribution. No information.

Medicinal use. In the Finschhafen area, the stem sap from this plant is drunk to give temporary relief to a bad cough. Soup made from aibika (*Abelmoschus manihot*) is usually given at the same time.

Chemistry. Alkaloids are not present (Hartley, 1973; Holdsworth, 1983).

Local name	Village name
Zanzanfulolo	Sosoningko

Callicarpa caudata Maxim. WEI-Bu-103, WEI-M-186

Description. Shrubs, with leaves, twigs and inflorescence all covered in felty hairs. The flowers are arranged in cymes, with the corolla a tube about 3 mm long with four stamens. The fruit is a succulent globose berry.

Ecology. Generally found in secondary or disturbed forest.

Distribution. No information.

Medicinal use. In a Morobe coastal village, the stem of this tree is scraped and mixed with the rhizome of ginger (*Zingiber officinale*) and applied to large wounds.

Chemistry. No information could be found.

Local name	Village name
no record	Buso

Premna obtusifolia R. Br.

Description. A small tree. The flowers consist of a short corolla tube with 4 stamens. The fruit is a fleshy, globose berry.

Ecology. Generally found in primary forest and on the beach.

Distribution. No information.

Medicinal use. In the coastal village of Buso, this small tree is burnt underneath the stilt houses to keep mosquitoes away.

In other parts of the country, uses include sniffing to clear nasal passages and rubbing on tropical ulcers in New Ireland, the treatment of headache or pneumonia on New Hannover Island, the treatment of headache or fever in North Solomons Province and the treatment of headache and intestinal worms in Milne Bay Province.

Chemistry. Traces of alkaloids have been extracted from the related species *P. integrifolia* and *P. corymbosa* (Hartley, 1973).

Local name	Village name
Acakaro	Buso

ZINGIBERACEAE

Amomum aculeatum Roxb.

(Gr. amomon – an ancient seasoning)

Description. Leafy shoots 2–4 m high. Leaves lanceolate, acuminate. Underground rhizome. Inflorescence with base in the ground. Flowers erect, corolla tube with transparent lobes, orange. Lip orange-yellow with crimson spots. Fruit covered with fleshy spines, in a dense short head on the thickened axis of the inflorescence. Fruit develops a mucilaginous mass from the decaying bracts. The fruit is edible.

Ecology. Found frequently in primary and secondary forest; also cultivated.

Distribution. Occurs in Java, Sumatra, Penang (Malaya).

Medicinal use. The sap obtained by beating the bark with a stick is given to a person with malaria or fever. The dose is about a cupful and may be repeated daily as necessary.

Chemistry. Alkaloids are not present (Holdsworth, 1983). Other species of *Amomum* contain indanecarboxaldehydes (Takido, 1978).

Local name	Village name
Galengapace	Sililio

Zingiber officinale Roxb.

(Fig. 47)
WEI-B-01, WEI-N-02, WEI-M-03

Description. Leafy stems to 50 cm high. Leaves dark green, long narrow blades. Rhizome either entirely pale yellowish or with a red external layer. Scape slender, to 12 cm tall. Corolla tube with yellow lobes. Rarely flowers (in Malaya).

Ecology. Occurs cultivated in tropical Asia from ancient times.

Distribution. The country of origin is unknown; now occurs cultivated throughout the tropics.

Medicinal use. There are different local races of this species and those with the red rhizomes are used medicinally. This plant is used throughout Morobe Province and the rest of Papua New Guinea. In Morobe Province alone, eight villages investigated were found to use this herb. The rhizome is chewed to treat stomach ache and diarrhoea in the villages of Biawen and Nauti and to relieve cough in Nasingalatu, Bolinbaneng, Sosoningko and Manki. Also in Manki, the rhizome and leaves are chewed and rubbed on the head to relieve headache. In Nasingalatu, the crushed rhizome is mixed with lime and left on a sore for three days. In Buang, the plant is ashed and placed on spear wounds which have been washed with water which has been run over heated quartz rock.

In other parts of the country, uses include the treatment of malarial fever (Madang and Central Province), coughs, influenza, sore throat, (in coastal central villages). The plant is also used as a contraceptive in Manus Province and records have been made in PNG of its treatment for sterility, pneumonia, intestinal worms, tuberculosis, aching teeth. In New Britain, it is recorded as being used to treat fever, body aches, rheumatism, toothache and tropical ulcers (Futscher 1959).

In Malaysia, the plant is used as an abortifacient (Burkill 1966) and in the West Indies, tea made from the rhizome is used to treat indigestion, stomach ache and malaria (Ayensu 1981).

Chemistry. Alkaloids are not present (Holdsworth, 1983). The major components of the essential oil are the sesquiterpenes A- and B-zingiberene and curcumene (Bauer, 1984). In the related species, *Z. zerumbet*, many sesquiterpenes have been obtained. These include humulenol 1 and 2, humulene epoxide 1 and 2 and zerumbone (Damodaran, 1968; Matthes, 1980). Monoterpenes, flavonoids and lignans are also present (Matthes, 1980).

Local names	Village names
Kovulu kirelhi	Biawen
Taua	Nauti
Asuvate	Manki
Laki	Nasingalatu
Taqe	Bolinbaneng, Sosoningko
Kanga	Aseki
Yasink	Buang

Fig. 47 *Zingiber officinale* Roxb.

APPENDIX

Definitions of botanical terms

Plant – General

deciduous	plants which shed all their leaves at one time of the year
decumbent	stems lying on the ground, but with their ends turned up
dioecious	a species in which the male and female flowers occur on separate plants
epiphytic	when a plant grows upon another and is not rooted in the soil
monoecious	a species in which both the male and female flowers occur on the same plant
procumbent	lying flat on the ground

Leaves

coriaceous	leathery in texture
glabrous	without hairs
pubescent	with hairs
rugose	wrinkled
sessile	without a petiole; leaf blade attached directly to the stem
stipules	a pair of appendages located at the base of the petiole where it joins the stem

Flowers

bisexual	a flower with both stamens (male) and carpels (female)
tepals	petal-like sepals, when there is no corolla
unisexual	a flower with either stamens or carpels but not both

Fruit

achene	when the (one) seed is tightly enclosed by the fruit wall (a dry fruit)
berry	a fleshy fruit, the outer skin and inner flesh soft
capsule	a dry, dehiscent fruit with 2 or more seeds
dehiscent	a fruit which opens when ripe (by sutures, pores or caps)

drupe	a fleshy fruit with soft outer skin, and inner flesh and a hard inner part around the seed
follicle	a dry fruit that dehisces along one suture
pappus	an outgrowth of hairs or bristles from the summit of the achene of the Compositae family
pod	a dry fruit that dehisces along two sutures; the fruit of the legume family

Information about flowers and leaves

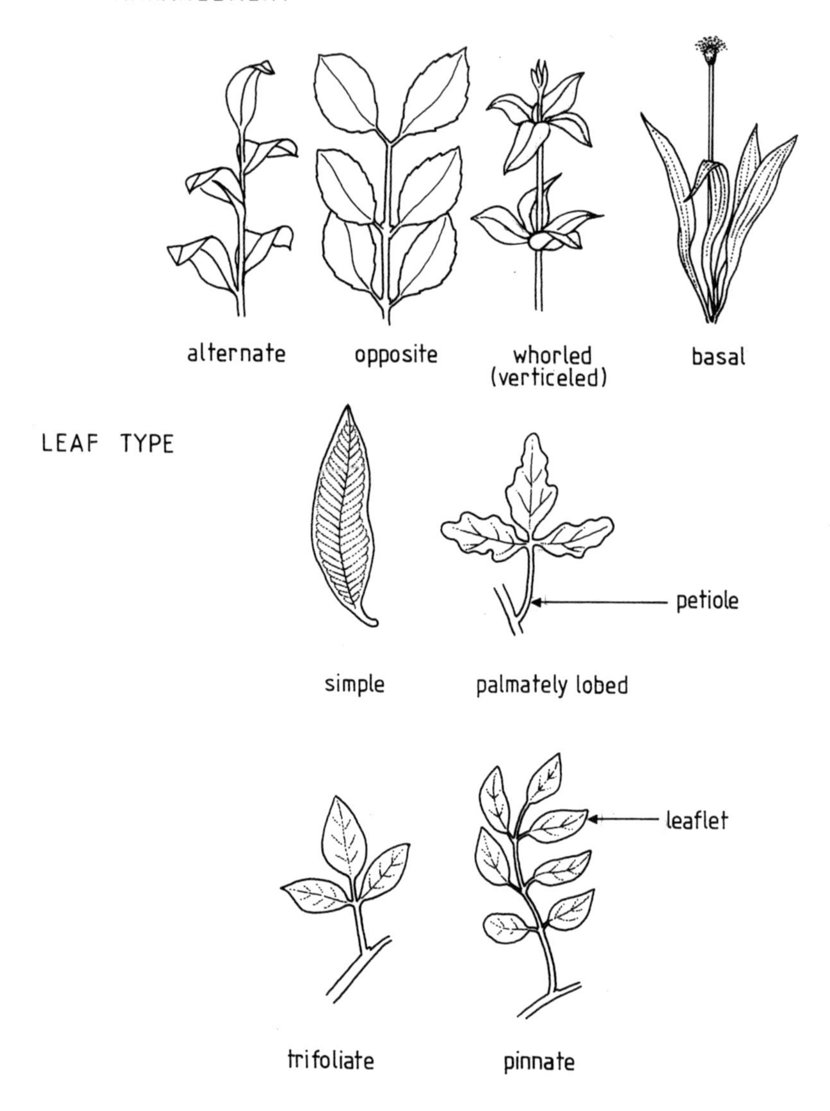

LEAF SHAPE

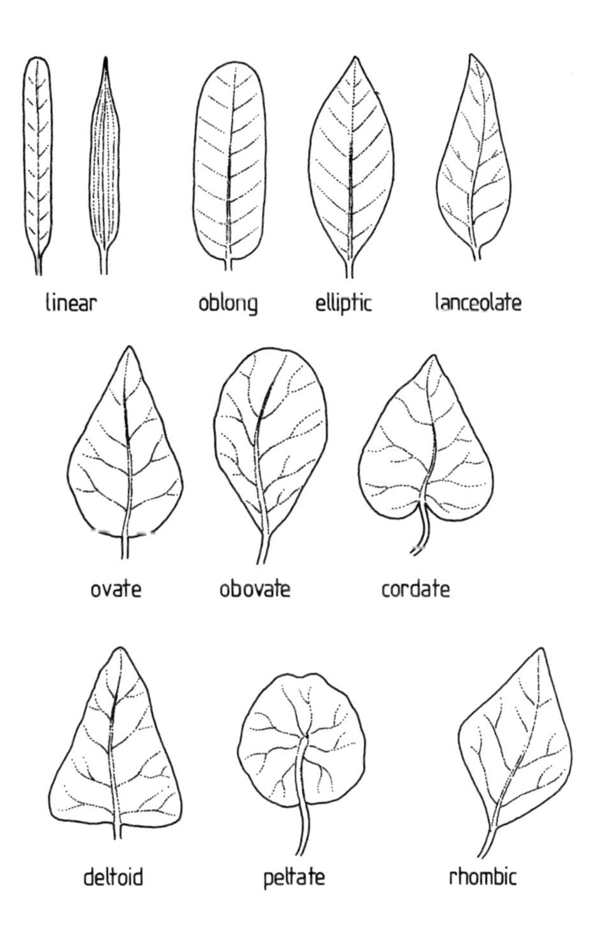
linear
oblong
elliptic
lanceolate
ovate
obovate
cordate
deltoid
peltate
rhombic

LEAF TIP

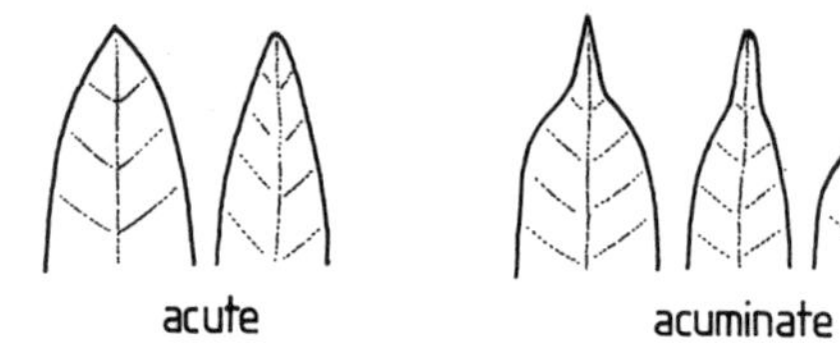

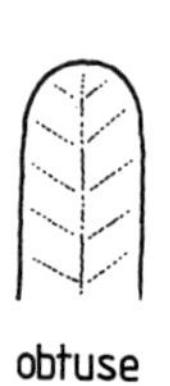

LEAF MARGIN

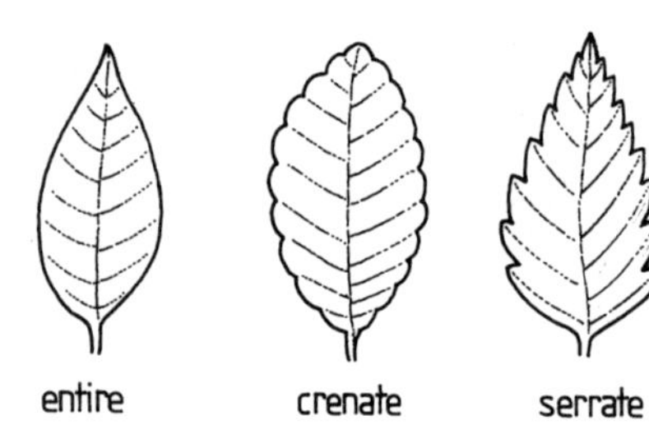

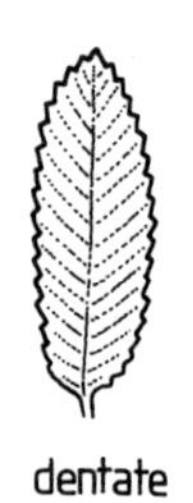

LEAF SURFACE

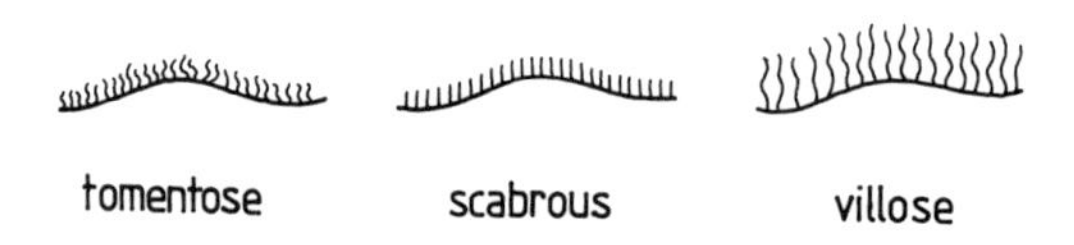

FLOWER ARRANGEMENT

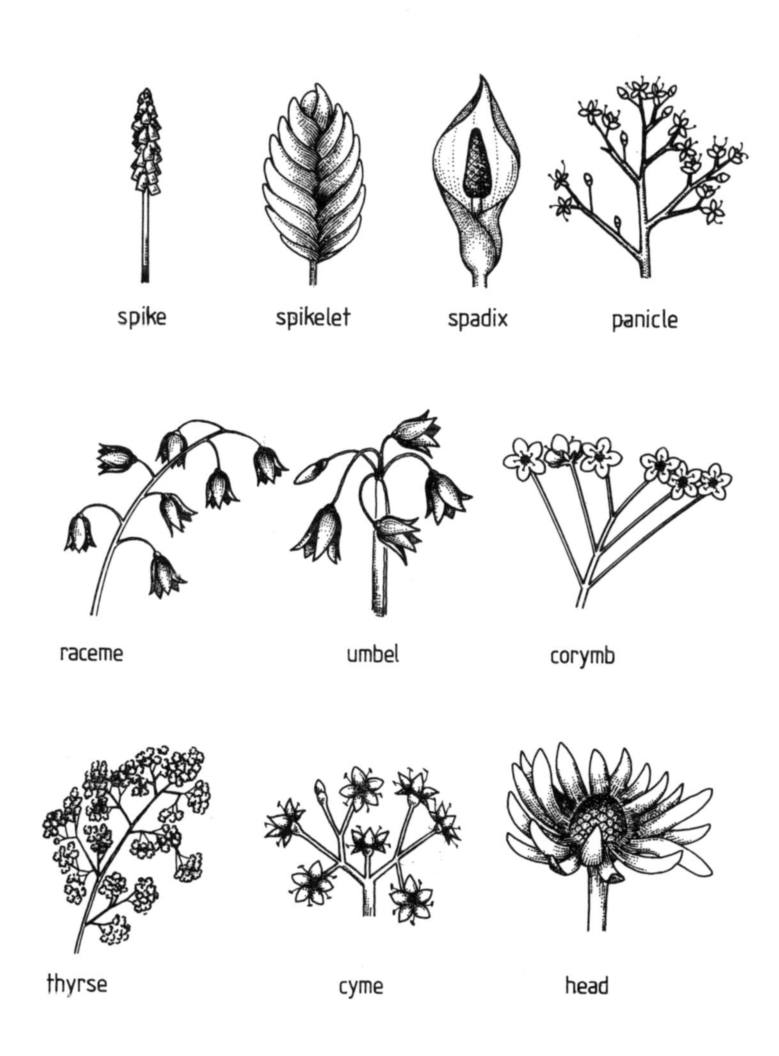

FLOWER STRUCTURE

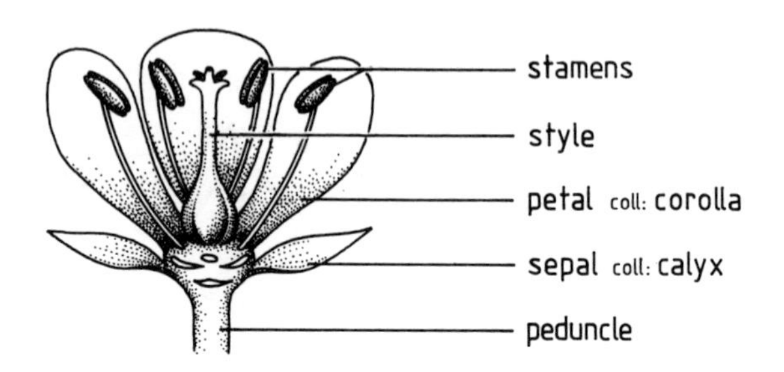

FLOWER LOCATION

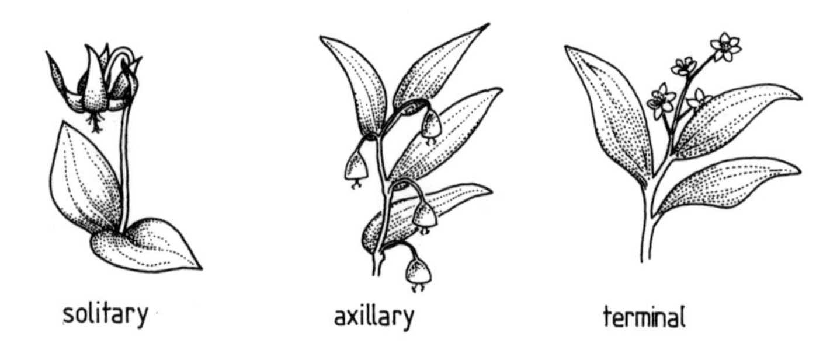

FLOWER SHAPE

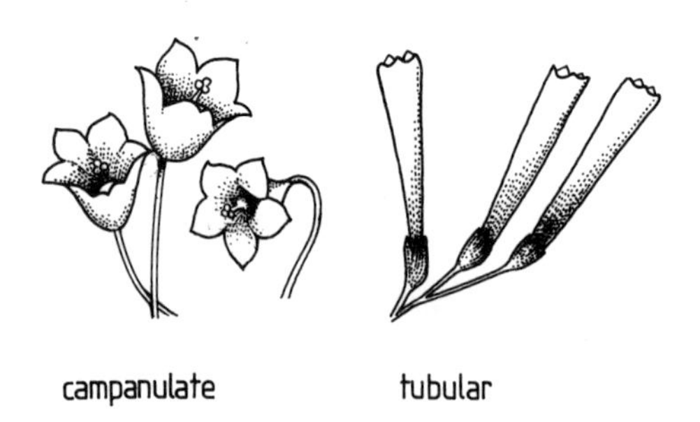

REFERENCES

BOTANICAL REFERENCES

The following references were used in compiling the botanical descriptions, ecology and distributions. Many of them are journals comprised of a number of volumes; they are not specifically referenced in the body of the text.

Airy Shaw, H. K. 1966. Notes on Melanesian and Other Asiatic Euphorbiaceae, Kew Bulletin 20: 25–49

Airy Shaw, H. K. 1974. Notes on Melanesian and Other Asiatic Euphorbiaceae, Kew Bulletin 29: 281–331.

Chew, W. L. 1969. *Nothocnide* (Urticaceae) in Malesia. The Gardens' Bulletin, Singapore 24: 361–373

Chew, W. L. 1969. A Monograph of *Laportea* (Urticaceae). The Gardens' Bulletin, Singapore 25: 111–178.

Chew, W. L. 1972. The Genus *Piper* (Piperaceae) in New Guinea Solomon Islands and Australia. Journal of the Arnold Arboretum 53: 1–25.

Corner, E. J. H. 1960. Taxonomic Notes on *Ficus* Linn., Asia and Australia. The Gardens' Bulletin, Singapore 17: 368–404

Corner, E. J. H. 1960. Taxonomic Notes on *Ficus* Linn., Asia and Australia. The Gardens' Bulletin, Singapore 17: 442–485

Corner, E. J. H. 1960. Taxonomic Notes on *Ficus* Linn., Asia and Australia. The Gardens' Bulletin, Singapore 18: 1–35

Corner, E. J. H. 1960. Taxonomic Notes on *Ficus* Linn., Asia and Australia. The Gardens' Bulletin, Singapore 18: 36–64

Corner, E. J. H. 1960. Taxonomic Notes on *Ficus* Linn., Asia and Australia. The Gardens' Bulletin, Singapore 18: 83–97

Corner, E. J. H. 1960. Taxonomic Notes on *Ficus* Linn., Asia and Hartley, T. G. 1974. A Revision of the Genus *Acronychia* (Rutaceae). Journal of the Arnold Arboretum 55: 469–567

Havel, J. J. 1975. Training Manual for the Forestry College Papua New Guinea Department of Forests, Port Moresby.

Henty, E. E. 1969. A Manual of the Grasses of New Guinea. Botanical Bulletin 1, Department of Forestry Port Moresby, PNG.

Holthius, L. B. and H. J. Lam 1942. Blumea 5(1): 220.

Huang, T. C. 1966. Monograph of *Daphniphyllum* II. Taiwania 12: 137–234.

Jayaweera, D. M. A. 1963. The Rubiaceous Genus *Mussaenda*: The Morphology of the Asiatic Species. Journal of the Arnold Arboretum 44: 111–126

Kalkman, C. 1953. Revision of the Burseraceae of the Malaysian Area in a Wider Sense. Blumea 7: 458–472.

Koster, J. Th. 1966. The Compositae of New Guinea. Nova Guinea 24

Koster, J. Th. 1970. The Compositae of New Guinea I. Blumea 18: 137–145

Koster, J. Th. 1972. The Compositae of New Guinea II. Blumea 20: 193–226

Koster, J. Th. 1976. The Compositae of New Guinea III. Blumea 23: 163–175

Koster, J. Th. 1979. The Compositae of New Guinea IV. Blumea 25: 249–282

Koster, J. Th. 1980. Blumea 26(1).

Kostermans, A. J. G. H. 1968. Materials for a Revision of Lauraceae I. Reinwardtia 4: 1–40.

Kostermans, A. J. G. H. 1969. Materials for a Revision of Lauraceae II. Reinwardtia 7: 451–536.

Kostermans, A. J. G. H. 1970. Materials for a Revision of Lauraceae I. Reinwardtia 8: 21–196.

Krukoff, B. A. 1972. Notes on the Asiatic-Polynesian-Australian Species of *Erythrina* II. Journal of the Arnold Arboretum 53: 128–139.

Leenhouts, P. W. 1973. A Revision of *Haplolobus* (Burseraceae). Blumea 20: 283–310.

Peekel, P. G. 1984. Flora of the Bismarck Archipelago for Naturalists. Office of Forests, Lae PNG.

Philipson, W. R. 1982. A Revision of the Malesian species of *Palmeria* (Monimiaceae). Blumea 28: 85–101.

Purseglove, J. W. 1974. *Tropical Crops,* Volumes I and II ELBS and Longman, England.

Randeria, A. J. 1960. Blumea 10(1)

Ridsdale, C. E. 1978. Blumea 24(2): 333–334.

Royen, P. van 1965. Manual of the Forest Trees of Papua and New Guinea, Part 5 – Himantandraceae, Division of Botany, Lae, PNG.

Royen, P. van 1965. Manual of the Forest Trees of Papua and New Guinea, Part 9 – Apocynaceae, Division of Botany, Lae, PNG.

Royen, P. van 1969. The Genus *Rubus* (Rosaceae) in New Guinea Phanerog. Monographs 2: 1–126.

Shaw, H. K. 1980. The Euphorbiaceae of New Guinea, Kew Bulletin Additional Series, Vol VIII, 243 pp.
Sleumer, H. 1960. Florae malesianae precursores 23. The Genus *Rhododendron* in Malaysia. Reinwardtia 5: 5–231.
Sleumer, H. 1966–7. Ericaceae. Flora Malesiana ser. I 6: 469
Streimann, H. 1983. The Plants of the upper Watut Watershed of Papua New Guinea. National Botanic Gardens, Canberra Australia.
Verdecourt, B. 1979. A Manual of the New Guinea Legumes. Botanical Bulletin II. Division of Botany, La.
Womersley, J. S. 1978. Handbooks of The Flora of Papua New Guinea. Melbourne University Press, 278 pp.
Zandee, M. and C. Kalkman 1981. The Genus *Rubus* (Rosaceae) in Malesia 1. Blumea 27: 75–113.

MEDICINAL REFERENCES

Ayensu, E. S. 1979. Plants for Medicinal Uses with Special Reference to Arid Zones. In: Arid Land Plant Resources Goodin, J. R. and D. K. Northington (eds), Smithsonian Institution, Washington, D.C.
Blackwood, B. 1935. Both Sides of the Buka Passage. Claredon Press, Oxford.
Burkhill, I. H. 1966. A Dictionary of the Economic Products of the Malay Peninsula. Ministry of Agriculture, Kuala Lumpur
Futscher, P. O. 1959. (as cited in Holdsworth, 1977)
Gideon, O. 1979. Medicinal Plants Study – Manus and New Ireland. Unpublished report, Hamilton, 1960.
Holdsworth, D. K. 1977. Medicinal Plants of Papua New Guinea. South Pacific Commission, Noumea.
Holdsworth, D. K. 1983. A Survey of Plants Used in Traditional Medicine on Karkar Island, Madang Province, Papua New Guinea. Science in New Guinea 9(3): 130–140.
Holdsworth, D. K. and P. Mahana 1982. A Phytochemical Survey of Medicinal Plants of the Morobe Province. Part I. Mountains of the Huon Peninsula. Science in New Guinea 9(1): 38–49.
Holdsworth, D. K and H. Sakulas. 1984. Plants and their Traditional Medicinal Uses. Aseki and Buang. Unpublished report.
Holdsworth, D. K. and H. Sakulas. 1986. Medicinal Plants of the Morobe Province. Part II. The Aseki Valley. International Journal of Crude Drug Research. Vol. 24

Holdsworth, D. K. and K. Damas. 1986. Medicinal Plants of Morobe Province, Papua New Guinea. Part III. The Finschhafen Coast. International Journal of Crude Drug Research, Padua et al. 1977.
Peekel, P. G. 1984. Flora of the Bismarck Archipelago for Naturalists. Office of Forests Division of Botany Lae, Papua New Guinea. Translated by E. E. Henty.
Perry, L. M. 1980. Medicinal Plants of East and Southeast Asia. Cambridge, Mass. USA. The MIT Press.
Powdermaker, H. 1933. Life in Lesu: the Study of a Melanesian Society in New Ireland. Williams and Norgate, London.
Quisumbing, E. 1951. Medicinal Plants of the Philippines. Philippines Department of Agriculture and Natural Resources Technical Bulletin 16: 1049–1078.
Rageau, J. 1973. Les Plantes Medicinales de la Nouvelle-Caledonie. Travaux et Documents de l'ORSTOM, Paris. Office de la Recherche Scientifique et Technique Outre-Mer.
Soepardi, R. 1967. (as cited in Holdsworth, 1977).
Steenis-Kruseman, M. J. van 1953. Select Indonesian Medicinal Plants. Bulletin No. 18. Organization for Scientific Research in Indonesia, Djakarta.
Sterly, J. 1970 (as cited in Holdsworth, 1977).
Stopp, K. 1963. Medicinal Plants of the Mount Hagen People (Mbowamb) in New Guinea. Economic Botany 17(1): 16–22.
Uhe, G. 1974. Medicinal Plants of Somoa. Economic Botany 28(1): 1–30.
Warburg, G. 1899 (as cited in Holdsworth, 1977)
Watt, J. M. and M. G. Breyer-Brandwijk, 1962. The Medicinal and Poisonous Plants of Southern and Eastern Africa. Livingstone, London.
Webb, L. J. 1948 (as cited in Holdsworth, 1977).
Webb, L. J. 1960. Some New Records of Medicinal Plants Used by the Aborigines of Tropical Queensland and New Guinea. Proceedings of the Royal Society of Queensland 71(16): 103–110.
Zepernick, B. 1972. (as cited in Holdsworth, 1977).

CHEMICAL REFERENCES

Abe, F. et al. 1977. Chemical Pharmaceutical Bulletin 25: 3422.
Achenback, H. et al. 1979. Tetrahedron Lett. 2571.
Adesogan, E. K. 1978. Journal of the Chemical Society and Chemical Community 152.

Arthur, H. R. et al, 1955. Journal of the Chemical Society 3740: 1960; 3197: 1961, 551.
Baslas, R. K. et al. 1980. Current Science 49(8): 311.
Bauer, K. and D. Garbe 1984. Common Fragrance and Flavour Materials. VCH, Germany.
Bechmann, S. et al. 1971. Phytochemistry 10: 2465.
Bhatia, V. K. et al. 1967. Tetrahedron 23: 1363.
Bhattacharya, A. K. et al. 1969. Journal of the Indian Chemical Society 46: 381.
Bick, I. R. C. et al. 1965. Australian Journal of Chemistry 18: 1997.
Blunden, G. et al. 1981. Journal of Natural Products 44 (4): 441.
Bohlmann, F. et al. 1979. Phytochemistry 18: 1062.
Bohlmann, F. et al. 1980. Phytochemistry 19: 2047.
Bohlmann, F. et al. 1981. Phytochemistry 20: 751.
Cavill, G. W. K. et al. 1968. Australian Journal of Chemistry 21: 2819.
Chemical Abstracts 1968. 69: 38737h.
Chemical Abstracts 1978. 89: 24706y.
Connell, D. W. et al. 1970. Australian Journal of Chemistry 23: 369.
Cooke, R. G. et al. 1964. Australian Journal of Chemistry 17: 379.
Corsano, S. et al. 1968. Journal of the Chemical Society and Chemical Community 738.
Damodaran, N. P. et al. 1968. Tetrahedron 24: 4123–4133.
Dan, S. 1980. Journal of Indian Chemical Society 57(7): 760.
Dawson, R. M. et al. 1966. Australian Journal of Chemistry 19: 2133.
Deshpande, V. H. et al. 1973. Indian Journal of Chemistry 11: 518.
Deshpande, V. H. et al. 1977. Indian Journal of Chemistry, Section B 158(3): 205.
Dictionary of Organic Compounds 1982. Chapman and Hall, New York.
Duke, J. A. 1985. CRC Handbook of Medicinal Herbs. CRC Press, USA.
Everist, S. L. 1974. Poisonous Plants of Australia. Angus and Robertson, Sydney.
Ewing, A. J. 1953. Australian Journal of Chemistry 6: 78.
Fagoonee, I. 1981. Insect Science: Its Applications 1 (14): 373.
Farnsworth, N. R. 1975. Journal of Pharmaceutical Sciences 64 (5): 717.
Farnsworth, N. R. 1980. Journal of Ethnopharmacology 2: 173.
Gautier, J. et al. 1972. Tetrahedron Letters 2715.
Gellert, E. et al. 1970. Australian Journal of Chemistry 23 : 2095.

Glosse, P. et al. 1978. Tetrahedron Letters 397.
Gough, J. et al. 1964. Australian Journal of Chemistry 17 (1270).
Govindachari, T. R. et al. 1956. Journal of the Chemical Society 629: 1957, 545, 548.
Govindachari, T. R. et al. 1965 Tetrahedron Letters 1907.
Govindachari, T. R. 1967. Tetrahedron 23: 1901.
Guenther, E. 1952. The Essential Oils. Van Nostrand Vol. 6.
Gui, Xiao-Ming, et al. 1980. Chung Ts'ao 11(5): 196.
Gupta, D. et al 1976. Indian Journal of Experimental Biology 14(3): 356.
Hart, N. K. et al. 1967. Chem. Commun. 87
Hartley, T. G. 1973. Lloydia 36(3): 217.
Haslam, E. et al. 1967. Journal of the Chemical Society (C) 2381.
Herbert, R. B. et al. 1972. Phytochemistry 11: 1184.
Hertz, W. et al. 1975. Phytochemistry 14:233.
Holdsworth, D. et al. 1983. International Journal of Crude Drug Research 21 (3): 121.
International Trade Centre, UNCTAD, GATT 1982. Markets for Selected Medicinal Plants and Their Derivatives Geneva.
Johns, S. R. et al. 1966. Australian Journal of Chemistry 19: 1951.
Johns, S. R. et al. 1967. Australian Journal of Chemistry 20: 1787.
Johns, S. R. et al. 1970. Australian Journal of Chemistry 23: 1119.
Johns, S. R. et al. 1971. Australian Journal of Chemistry 24: 1679.
Kasturi, T. R. et al. 1973. Indian Journal of Chemistry 11: 91.
Krishnamurty, H. et al. 1980. Phytochemistry 19: 2797.
Krishnaswamy, M. et al. 1980. Indian Journal of Experimental Biology 18(11): 1359.
Kuchl, F. A. et al. 1948. Journal of the American Chemical Society 70: 2091.
Lamberton, J. A. et al 1953. Australian Journal of Chemistry 6:66.
Lassack, E. V. 1983. Australian Medicinal Plants. Methuen Australia
Lasswell, W. et al. 1977. Phytochemistry 16: 1439.
Manske, R. H. F. 1960. The Alkaloids, Vol. 6. Academic Press New York.
Manske, R. H. F. 1965. The Alkaloids, Vol. 8. Academic Press New York.
Manske, R. H. F. 1968 The Alkaloids, Vol. 11. Academic Press New York.
Manske, R. H. F. 1970. The Alkaloids, Vol. 12. Academic Press New York.

Manske, R. H. F. 1975. The Alkaloids, Vol. 14. Academic Press New York.
Manske, R. H. F. 1975. The Alkaloids, Vol. 15. Academic Press New York.
Manske, R. H. F. 1979. The Alkaloids, Vol. 17. Academic Press New York.
Manske, R. H. F. 1981. The Alkaloids, Vol. 18. Academic Press New York.
Manske, R. H. F. 1981. The Alkaloids, Vol. 19. Academic Press New York.
Manske, R. H. F. 1985. The Alkaloids, Vol. 25. Academic Press New York.
Mathieson, D. W. et al. 1973. Journal of the Chemical Society Perkin Trans., 184.
Matthes, H. et al. 1980. Phytochemistry 19: 2643.
Mujumdar, A. M. et al. 1981. Journal of Plant Crops 7 (2): 69
Nahrstedt, A. et al. 1982. Phytochemistry 20(11): 2587.
Nair, A. G. et al. 1977. Ind. Journal of Pharmacology 39 (5): 108.
Nicollier, G. et al. 1981. Phytochemistry 20(11): 2587.
Oelrichs, P. B. et al. 1980. Journal of Natural Products 43(3): 414.
Perry, L. M. 1980. Medicinal Plants of East and Southeast Asia. MIT Press, USA.
Polansky, J. 1957. Bull. Soc. Chim. Fr. 1079.
Polansky, J. 1958. Bull. Soc. Chim. Fr. 929.
Polansky, J. 1969. Phytochemistry 8: 1221.
Rehfeldt, A. et al. 1980. Phytochemistry 19(8): 1685.
Ritchie, E. and W. C. Taylor 1967. The *Galbulimima* Alkaloids. In: Manske, R. H. F. The Alkaloids, Vol IX.
Row, I. et al. 1962. Tetrahedron 18: 827.
Satake, T. 1980. Chemical Abstracts 93: 128747 m.
Saxena, V. K. et al. 1976. Planta Medica 29: 96.
Schneider, W. P . et al. 1965. Journal of Organic Chemistry 30: 2856.
Simes, J. J. H. et al. 1959. Australian Phytochemical Survey Part 3, CSIRO Bulletin No. 281. Government Printer, Melbourne.
Singh, M. et al. 1982. Planta Med. 44(3): 171.
Sung, Chung-Ki et al. 1980. Journal of the Chemical Society, Chemical Commun. 19: 909.
Tagagi, S. et al. 1977. Yakugaku Zasshi 97(10): 1155.
Tutin, F. et al. 1911. Journal of the Chemical Society 99: 946.
Takeda, Y. et al. 1977. Phytochemistry 16: 1401.
Takido, M. et al. 1978. Phytochemistry 17: 327.

Webb, L. J. 1948. Guide to the Medicinal and Poisonous Plants of Queensland. CSIRO Bulletin No. 232. Government Printer Melbourne.
Webb, L. J. 1959. Proceedings of the Royal Society of Queensland 71: 103.
Xiao, P. G. 1981. Fitoterapia 52: 65.
Yamamura, S. et al. 1977. Bulletin of the Chemical Society of Japan 50(7): 1836.

INDEX

WAU ECOLOGY INSTITUTE HANDBOOK No. 10

SUBSISTENCE AGRICULTURE IMPROVEMENT

Manual for the Humid Tropics

Edited by
Friedhelm Goeltenboth